Heinemann Experimental Chemistry Series

General Editors:

A. J. Mee
Martin Rogers, *Headmaster, Malvern College*

Advisory Panel:

Professor E. H. Coulson, *Organizer, Nuffield Advanced Chemistry Project*
Professor H. F. Halliwell

2
Electrochemistry

Students' Text

Heinemann Experimental Chemistry Series

1—Structure, Equilibrium, and Kinetics
A. J. Mee and Derek Bradford

***2—Electrochemistry**
D. Mansfield

***3—Chemical Analysis, Chromatography, and Ion Exchange**
R. T. Allsop and J. A. D. Healey

4—Transition Element Chemistry
J. N. King and T. McManus

5—Atomic and Molecular Masses
Derek Bradford

6—Colloids
D. Mansfield

7—Carbon Chemistry
A. G. Rayden

8—The Periodic Table
Claude Evans

***9—Metals and Alloys**
F. W. J. Pargeter and M. R. Mander

***10—Radiochemistry**
D. Mansfield

Note: only those titles marked * have been published at the date of publication of the present volume.

Heinemann Experimental Chemistry Series

2
Electrochemistry

Students' Text

D. Mansfield, B.Sc.

Head of Chemistry Department
Harvey Grammar School, Folkestone, Kent

Heinemann Educational Books

Heinemann Educational Books Ltd

London Edinburgh Melbourne Auckland Toronto
Hong Kong Singapore Kuala Lumpur
Ibadan Nairobi Johannesburg
Lusaka New Delhi

ISBN 0 435 65952 9

First published 1974

Published by Heinemann Educational Books Ltd
48 Charles Street, London W1X 8AH

Printed in Great Britain by
William Clowes & Sons Limited
London, Beccles and Colchester

Preface to the Series

The richness of the British tradition of science teaching lies in its experimental work. Nowhere in the world has there been for so long such an emphasis on experiments performed both by teachers and students. The books in this series have been written to sustain that tradition and to inject new life into it.

Much of the practical work in chemistry which was done by students in the past consisted merely of the repetition of experiments, the results of which were already known, or in the preparation of compounds by following a rigid set of instructions. In either case there was little room for thought on the part of the student. Modern chemistry teaching seeks to improve on this situation by emphasizing that the aim of practical work is to discover something, not merely to copy what has been done before. Of course, experimental work at this level is not, in fact, research, but there is no reason why it should not be presented as if it were. To the student it *is* something new that he is finding out.

These books have been designed to cover the whole area of advanced school chemistry and introductory courses at colleges and universities. A series of books, each centred on a specified topic, offers considerable advantages over a single volume. It is possible to include more material and to present it in a very flexible way. Teachers can choose those volumes which suit their own particular course and students are not stuck with a heavy volume containing a good deal of irrelevant material. Moreover, each book can be written by a teacher, expert in a particular field.

What is new in this series is the method of treatment. The approach is based on discovery by the students. They are given sufficient information to perform the experiments in an investigational manner and to think about them, but not so much that the performance becomes a meaningless ritual. The experimental work is not an end in itself. The experiments serve some definite purpose in relation to the theory work. By means of questions students are led to examine the results of their work, not only with regard to the degree of accuracy they may have attained but to deduce theoretical points arising from the experiments. In this way the practical work becomes an integral part of the chemistry course and not a mere adjunct to it.

We have attempted to include a number of modern experiments which are new to the school laboratory. Teachers will recognize the many 'classical' experiments which also appear, although these have been treated with a different emphasis.

The books are intended to allow students to proceed at their own pace. This does not mean, of course, that the teacher is absolved from the responsibility of over-seeing the student's work, but it does mean that students can get on without having a lecture period beforehand in which the teacher explains what has to be done in the practical session. It is hoped that the use of the books will foster independence of thought among students and will thus enable the teacher to assume his proper role of counsellor and guide rather than instructor.

For each *Students' Text* a *Teachers' Guide* has been prepared. It gives hints about the conduct of the experiments and the conclusions that can be drawn from them, and contains the answers to most of the questions posed in the *Students' Texts*. The list of materials required for each experiment is provided in the *Students' Texts*.

We cannot hope to have produced a series of books which is completely free from error and we hope that users will inform us of any lapses. It would be foolish to claim, too, that the experiments described are incapable of improvement and we should be glad to hear of any way in which this can be done.

Specific acknowledgment of the considerable amount of help received in the writing of these books will be found in the Introduction to the individual volumes. We are, however, deeply indebted to Prof. E. H. Coulson and Prof. H. F. Halliwell for acting as consultants in the preparation of the series as a whole. They gave very valuable advice from which all the books have benefited.

The study of chemistry through experiments is not only great fun, it is an excellent education. We hope that the teaching of experimental chemistry will be encouraged and enriched by this series.

A. J. Mee
1974 Martin Rogers

Contents

Introduction

'Although we know nothing of what an atom is, yet we cannot resist forming some idea of a small particle, which represents it to the mind: and though we are in equal, if not greater, ignorance of electricity, so as to be unable to say whether it is a particular matter or matters or mere motion of ordinary matter, or some kind of third power or agent, yet there is an immensity of facts which justifies us in believing that the atoms of matter are in some way endowed or associated with electrical powers, to which they owe their most striking qualities, and amongst them their mutual chemical affinity.'

Michael Faraday, writing in *Experimental Researches in Electricity*

Faraday's classical researches in electrochemistry laid the foundations of this branch of science and, as the passage quoted above shows, suggested to him the ultimate electrical nature of matter. This, and later work in electrochemistry, has been arguably the most powerful single tool to contribute to the development of chemical theory in the last 150 years. Its contributions to commercial developments have been similarly striking—on a scale often unperceived by the original researchers. It is, however, recorded that Gladstone remarked to Faraday at the Royal Institution, 'This invention of yours is very interesting, Professor Faraday, but what is the use of it?' To this Faraday replied, 'Sir, there is every probability that in a few years' time you will be able to tax it!'

1: Equilibrium

1.1 Is there an Equilibrium between a Metal and a Solution of its Ions?

You may have been told in earlier work, when you were considering the electrochemical series, that there is an equilibrium between a metal and a solution of its ions. Thus, if a piece of silver is placed in a solution of silver nitrate, atoms of silver go into solution as ions at just the same rate as ions of silver come out of solution and attach themselves as atoms of silver to the silver plate.

Such a theory is useful in explaining such phenomena as electrode potentials. But is it true? What evidence is there that such dynamic equilibria occur?

Fortunately we now have a method of finding out whether such equilibria do exist by using radioactively 'tagged' atoms and ions. If there is an equilibrium

$$Ag(s) + aq \rightleftharpoons Ag^{+}(aq) + e^{-}$$

and a piece of non-radioactive silver foil is placed in a solution containing some radioactive silver ions, it would be expected that some of the non-radioactive silver atoms would be exchanged for radioactive ones.

^{110}Ag is radioactive. It emits β-particles and γ-radiation and has a half-life of 254 days. You will be provided with a solution of silver nitrate containing some ^{110}Ag ions.

This is probably the first time you have used radioactive materials. ***You must observe all the safety rules governing the use of these substances. These are described in Book 10 of this series.***

In order to detect radioactivity you will be using a Geiger–Müller counter (abbreviated to 'a G–M tube'). You need not know at this stage how it works as you will be using it purely as a tool. The G–M tube is connected to an instrument called a scaler which measures the number of particles radiated.

Apparatus and Materials

Piece of thin silver foil (1 × 1 cm)
Silver nitrate solution (about 0.1 M) containing some $^{110}Ag^{+}$
Small specimen tube with stopper
Infra-red drying lamp
G–M tube in holder, lead, and scaler
Stop-clock
Wash-bottle of distilled water
Usual protective accessories for radiochemical work

Procedure

Devise a method of holding the piece of silver foil in a fixed position relative to the G–M tube. Find the background radiation over five minutes with the foil in this position. Place some of the radioactive silver nitrate solution into the specimen tube and drop the piece of foil into it. Leave for one hour; then remove the foil, wash it with distilled water, and dry it under the infra-red lamp. Replace the silver foil in position in front of the G–M tube and recount for five minutes. What evidence does your result offer concerning the possibility of the establishment of the equilibrium?

Questions

Does your result show that equilibrium has been established or only that exchange has taken place? What modification to the experiment would you suggest to find this out?

2: Electrical Potential

2.1 How does the Voltage of a Daniell Cell vary with the Current drawn from the Cell ?

Chemical energy can be changed into other forms of energy. In a cell, chemical energy is converted directly, and in part, into electrical energy.

The electrical potential difference (sometimes called the voltage) between two charged bodies is the difference in level of the electrical energy contained by the two bodies. If the charged bodies are connected by a conductor electrical energy will flow along the conductor until the potentials are equalized. This is analogous to the flow of heat from a hot body to a cold body until the temperatures are the same.

A body has an electrical charge when it has an excess or a deficiency of electrons. In the former case it is negatively charged; in the latter positively. The movement of electrons along a conductor constitutes an electric current. The number of electrons passing a given point in the conductor in a certain time gives the quantity of charge passing and is measured in coulombs (ampere seconds). The greater the potential difference between the ends of a conductor the larger is the current that flows through it; for many types of conductor the current is, in fact, proportional to the potential difference (Ohm's law).

If a potential difference is to be maintained between two charged bodies connected by a conductor, thus giving rise to a continuous flow of current, it follows that energy must be supplied. This comes from a chemical reaction (as in a cell), or by the performance of mechanical work (as in a generator), or from heat (as in a thermocouple), or from light (as in a photo cell).

In Experiment 1.1 you have found that when a metal is immersed in a solution of its ions an equilibrium exists between the metal and its ions. In a Daniell cell a copper plate is immersed in a solution containing $Cu^{2+}(aq)$ and a zinc plate in a solution containing $Zn^{2+}(aq)$. Two equilibria are therefore possible

$$Cu^{2+}(aq) + 2e \rightleftharpoons Cu(s) \qquad (1)$$

and

$$Zn^{2+}(aq) + 2e \rightleftharpoons Zn(s) \qquad (2)$$

Equilibrium (1) lies further to the right than does equilibrium (2). In other words zinc has a greater tendency to form ions in aqueous solution than does copper. If the two reactions are coupled together we get the equilibrium

$$Cu^{2+}(aq) + Zn(s) \rightleftharpoons Zn^{2+}(aq) + Cu(s) \qquad (3)$$

which, because of the individual equilibria (1) and (2), lies very far to the right; so much so that to all intents and purposes it goes to completion.

The reaction

$$Cu^{2+}(aq) + Zn(s) \longrightarrow Zn^{2+}(aq) + Cu(s)$$

is a spontaneous one and must therefore be associated with a decrease in free energy of the system (see Book 1). When such reactions are carried out in electrolytic cells, a variable proportion of the total energy—up to a definite maximum—is obtainable as free energy and the remainder is produced as heat.

Apparatus and Materials

Small specimen tube with cork (or bung) in which a hole has been bored to take glass tubing of approximately 8 mm external diameter.
Glass tube about 10 cm long and 8 mm external diameter
High resistance voltmeter (0–1.5 V)
Milliammeter (0–1 mA)
Connecting leads and crocodile clips
Rheostat (2 kΩ)
Teat pipette
Polyfilla
Strip of zinc foil and strip of copper foil (about 10 cm × 0.5 cm)
1.0 M solutions of copper(II) sulphate and zinc sulphate
10 per cent solution of potassium nitrate

Procedure

Construct a Daniell cell, as in Figure 2.1, in the following manner. Mix sufficient Polyfilla using potassium nitrate solution instead of water. Make a plug at one end of the glass tube with the paste about 1 cm in depth and leave aside for several minutes until the Polyfilla has set. Pour copper(II)

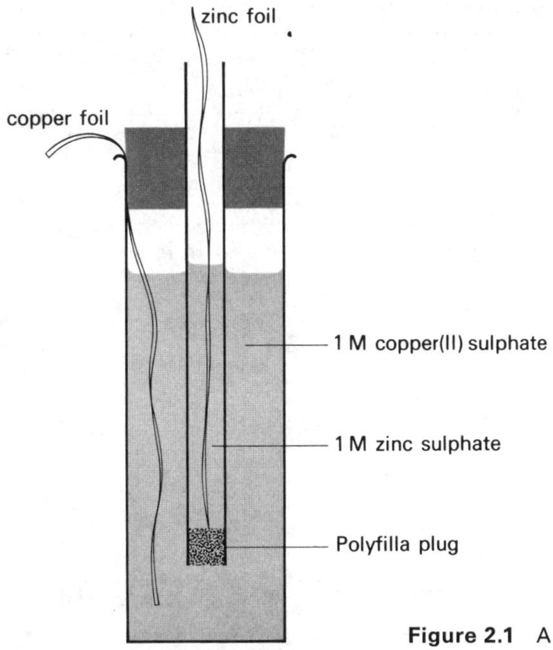

Figure 2.1 A Daniell cell

sulphate solution into the specimen tube to about two-thirds of its depth and fix the glass tube into this with the plug at the lower end until this almost reaches the bottom of the specimen tube. Use the teat pipette to add the zinc sulphate solution to the inside of the glass tube until the heights of the two liquids are approximately equal. Put the zinc foil into this solution.

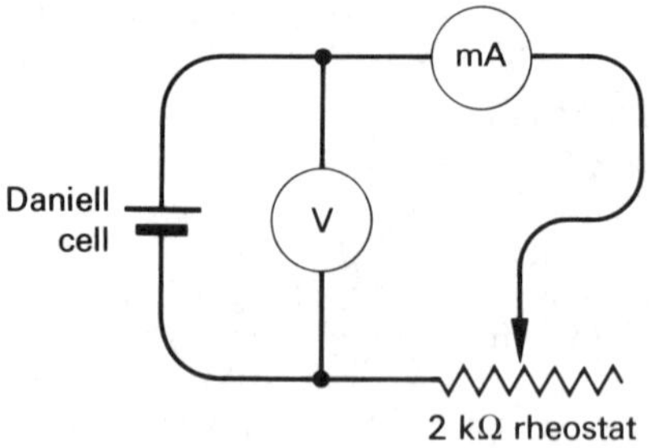

Figure 2.2 The experimental circuit

Assemble the circuit shown in Figure 2.2. Set the rheostat to minimum resistance and measure the voltage and current. Adjust the resistance until the potential difference is about 0.1 volt greater and measure both voltage and current once more. Continue to take readings in this way until the maximum resistance is included in the circuit.

Results

Record your results in a table as shown below.

Potential difference of cell in volts	*Current flowing in mA*	*Power in watts*

Plot your results, drawing two curves as shown in Figure 2.3.

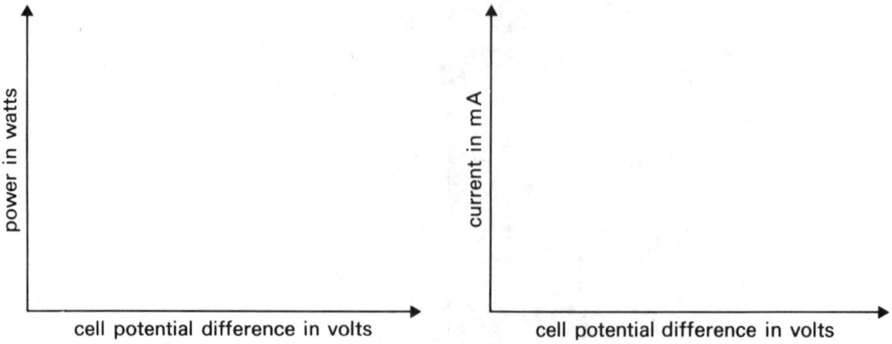

Figure 2.3

Questions

1. At what operating voltage is maximum power obtained from the cell?
2. What is the maximum potential obtainable from the cell? Under what conditions would this e.m.f. be obtained?
3. Can a cell be said to have an e.m.f. when it is not supplying current? How does this agree with the idea of thermodynamic reversibility? (See Book 1.)

Cell Notation

To avoid drawing such electrolytic cells (or making wordy descriptions of them) a standard notation is accepted. That for the Daniell cell is:

$$Zn(s)\,|\,Zn^{2+}(aq)\,\vdots\,Cu^{2+}(aq)\,|\,Cu(s) \qquad E = +1.1 \text{ volts}$$

The continuous lines in the notation represent breaks between phases, the dotted line represents the ionic bridge separating the two electrodes. The *sign* of E is, conventionally, the polarity of the right-hand side electrode.

Methods of Measuring Cell Potentials

As we have seen, the maximum cell potential exists when zero current is drawn from it, i.e. when the resistance of the external circuit is infinite. If we are to measure such cell potentials with accuracy, it is essential that we approach as closely as possible to these conditions. The following methods can be used in this connection.

(1) *Using a high-resistance voltmeter.* These have input resistances of up to 10^5 ohms and are thus fairly satisfactory.

(2) *Valve voltmeters (sometimes called d.c. amplifiers).* These often contain a special 'electrometer' valve as initial input stage with a resistance of 10^{14} ohms. More recent models may instead use a field-effect transistor in this stage and these have resistances of some 10^9 ohms.

(3) *Potentiometers*

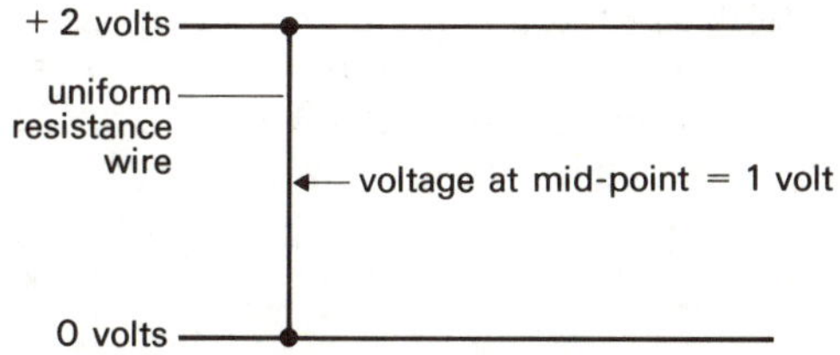

The parallel horizontal lines represent wires maintained at +2 volts and 0 volts respectively. They are joined by the uniform wire down which the voltage drops uniformly, thus at the mid-point it is +1 volt.

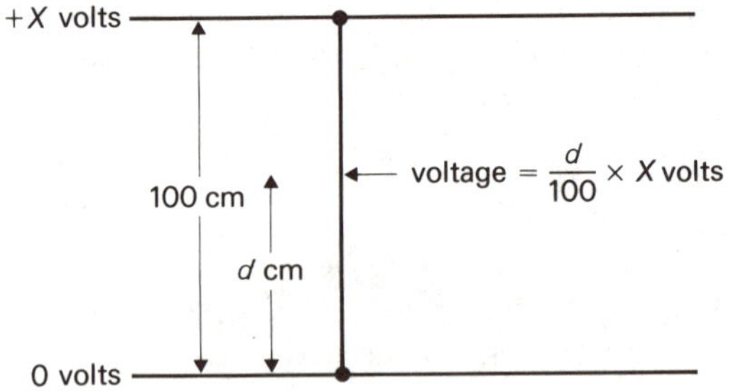

Here the general case is represented, if the potential difference between the parallel conductors is X volts and the resistance wire is 100 cm long. There is thus a voltage drop of (X/100)ths of a volt for every cm down the wire. This device can thus be used to divide a potential difference into any required fraction. It is known as a potential divider or potentiometer.

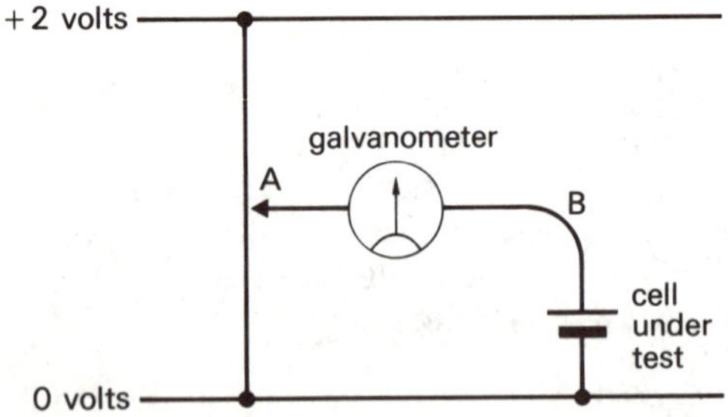

The diagram above shows how the potentiometer principle can be used to measure cell potentials. If the position of A is adjusted until the meter reading is zero, the potentials of A and B must be equal. If the potential at A is calculated as previously, this tells us the cell potential. Note that at the 'zero-point' no current flows from the cell, thus fulfilling the required conditions.

If you use this method of determining cell potentials you may be supplied with a potentiometer in which the resistance wire is stretched across a meter rule or with a more sophisticated version in which the resistance wire is coiled and lengths taken from it by a sliding contact. The volume controls on radios and record players are potentiometers which work in this way.

2.2 How does the Potential of an Electrolytic Cell vary with Temperature?

In the next few experiments you will investigate the effect on the potential of a cell of changing the possible variables. The cell chosen for study is

$$Cu(s)|Cu^{2+}(aq) \vdots Ag^{+}(aq)|Ag(s)$$

which is similar to the Daniell cell, but uses a silver electrode instead of a zinc electrode.

Which variables are likely to affect the potential of the cell? Would you expect temperature to be one of them? If so, why? Is the concentration of the ions likely to affect the potential? What effect is the area of the electrodes likely to have?

As we shall be concerned with the energy which is associated with the reaction taking place in the cell

$$Cu(s) + 2Ag^{+}(aq) \longrightarrow Cu^{2+}(aq) + 2Ag(s)$$

we need to know how this is related to the e.m.f. of the cell. When one mole of copper reacts, two moles of electrons are transferred from copper to silver ions, and two moles of solid silver are deposited. The charge transferred in this process is thus two faradays ($= 2 \times 96\,500$ coulombs). The volt is defined as one joule coulomb^{-1}, so the number of joules involved in this transfer is

$$-E_{\mathrm{T}} \times 2 \times 96\,500$$

where E_{T} is the e.m.f. of the cell at temperature T.

This energy is more conveniently expressed in kilojoules and amounts to

$$-E_{\mathrm{T}} \times 2 \times 96.5 \text{ kJ mol}^{-1}$$

The negative sign is used because the cell does electrical work and the energy content of the system therefore decreases (see Book 1).

If the e.m.f. E_{T} is measured when the current drawn is zero, we have a thermodynamically reversible system (see Experiment 2.1 and Book 1) and the energy calculated above is the maximum free energy obtainable from the system. It is therefore the ΔG of the equation

$$\Delta G = \Delta H - T\Delta S$$

where ΔH is the enthalpy change and ΔS is the entropy change (Book 1).

In the general case

$$\Delta G = -zFE$$

where F is the faraday (96 500 coulombs), z is the number of moles of electrons transferred when the reaction occurs, and E is the e.m.f. of the cell. The units are J mol^{-1}. Generally this is per mole of reactants, but in doubtful cases the reactant concerned should be specified. Thus, there is no dubiety in the case of the reaction occurring in the Daniell cell

$$Zn(s) + Cu^{2+}(aq) \longrightarrow Zn^{2+}(aq) + Cu(s)$$

because molar quantities react with each other; but in the cell to be used in this experiment the reaction takes place between one mole of copper and two moles of silver ions, and the ΔG calculated is per mole of copper.

Apparatus and Materials

U-tube with both limbs fitted with rubber stoppers each holding thermometer (−10 – 110 °C)
Strips of copper and silver foil to be used as electrodes
Connecting leads
Valve voltmeter or potentiometer
Polyfilla
10 per cent solution of potassium nitrate
1 dm^3 beaker, tripod, gauze, Bunsen burner
1.00 M solutions of copper(II) sulphate and silver nitrate

Procedure

Mix the Polyfilla with potassium nitrate solution as in Experiment 2.1 and assemble the cell as shown in Figure 2.4.

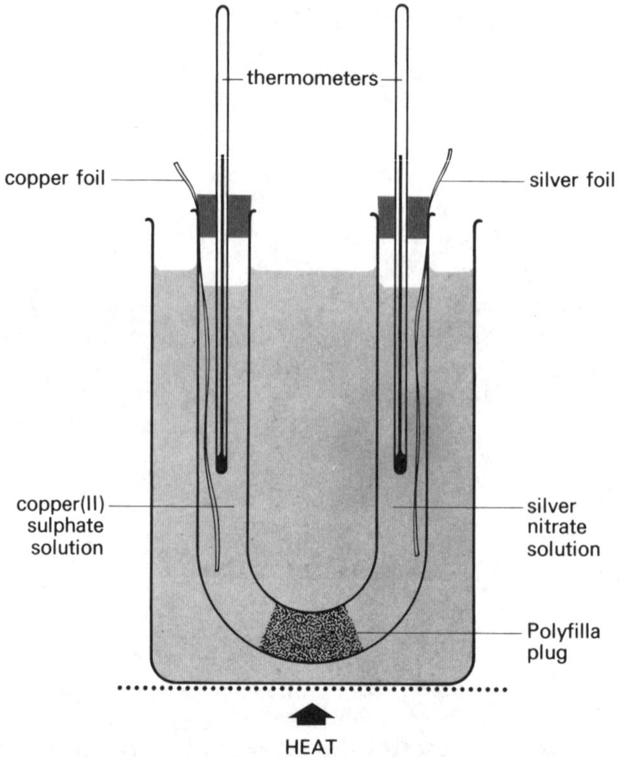

Figure 2.4 Effect of temperature on a cell potential

Note the temperature of the electrolytes and measure the cell potential (the temperatures should be the same but if they are not, take the average).

Raise the temperature by about 10 °C and again measure the cell potential. The temperature should be maintained as steady as possible while the potential is being taken.

Repeat, taking the cell potential at intervals of about 10 °C until about six readings have been obtained.

Recording and calculation

Record your results in a table as follows.

Temperature in K	*Cell potential in volts*	ΔG *in kJ mol*$^{-1}$

Plot ΔG against T and obtain the value of ΔG_{298}.

Discussion

The experiment shows that E, and therefore ΔG, is temperature dependent. It will be remembered that

$$\Delta G = \Delta H - T\,\Delta S$$

Since ΔH and ΔS are almost independent of temperature, what should be the form of the graph of ΔG against T? Do your results agree with this conclusion?

From the curve obtain values for ΔS and ΔH. Compare the value of ΔH with that obtained calorimetrically in Book 1.

Because the potentials are measured over a comparatively small range of temperature, extrapolation cannot be carried out with great accuracy and it would be more satisfactory, although rather time consuming, to calculate ΔH and ΔS from the results arithmetically.

Standard Free Energy Change

$\Delta G^{\ominus}_{298}$ was defined in Book 1 as the maximum free energy change of a reaction when the reactants, each at unit concentration, are converted to products, each at unit concentration, at the standard temperature of 298 K.

In this experiment we are dealing with reactants and products at unit concentration (the Ag^{+} and the Cu^{2+} solutions were molar) and we are working under thermodynamically reversible conditions. Hence the value of ΔG_{298} obtained above is the standard free energy change $\Delta G^{\ominus}_{298}$.

2.3 How does the Potential of a Cell vary with the Concentration of the Electrolytes?

The second variable that may affect the e.m.f. of a cell is the concentration of the electrolytes. We shall study the same cell as in Experiment 2.2. There are two electrolytes, Cu^{2+} and Ag^{+}, and it would be possible to examine the effect of changes of concentration of both. We shall, however, maintain the concentration of Cu^{2+} constant at 1.00 M and find the effect on the e.m.f. of the cell of changing the concentration of Ag^{+}.

What else must be kept constant during this experiment?

The experiment is best carried out on a class co-operative basis, each group constructing one or two cells, which are labelled and then used by other groups. In this way a number of results will be obtained for each concentration which can be averaged and used by the class as a whole.

Apparatus and Materials

Specimen tube with inner glass tube as for Experiment 2.1
Graduated pipette for making dilutions
Distilled water
Strips of copper and silver foil for electrodes
Valve voltmeter or potentiometer
Polyfilla
10 per cent solution of potassium nitrate
1.00 M solution of copper(II) sulphate
0.01 M solution of silver nitrate

Procedure

Using the general technique for constructing a cell described in Experiment 2.1 construct a series of cells each having 1.00 M copper(II) sulphate in the outer compartment with a copper foil electrode, and varying concentrations of silver nitrate with a silver foil electrode in the inner compartment. Suitable Ag^{+} concentrations are 0.01, 0.0033, 0.001, 0.000 33, and 0.0001 M. These should be prepared by progressive dilution.

Measure the e.m.f. of the cells.

Recording and Calculation

Record your results in a table as follows.

$[Ag^{+}(aq)]$ in mol dm^{-3}	$\log\,[Ag^{+}(aq)]$	E in volts

Include the result of Experiment 2.2.

Note that if the cell is written as

$$Cu(s)\,|\,Cu^{2+}(aq)\;\vdots\;Ag^{+}(aq)\,|\,Ag(s)$$

the sign of the e.m.f. is the polarity of the right-hand electrode, i.e. Ag. Plot a graph of log $[Ag^{+}(aq)]$ against E.

(a) If $E^{\ominus}$ is the e.m.f. of the cell when $[Ag^{+}(aq)] = 1$, obtain from the slope of the curve the value of k in the equation

$$E = E^{\ominus} + k \log[Ag^{+}(aq)]$$

(b) The intercept on the x axis gives the logarithm of the concentration of Ag^{+} ions for which the e.m.f. of the cell is zero, i.e. the logarithm of the concentration of Ag^{+} ions which would give a silver electrode potential (but opposite in sign) to that of a copper electrode for which the concentration of Cu^{2+} ions is 1. In such a cell no electrons would flow as the e.m.f. is zero. Hence these concentrations of ions may be said to be in equilibrium. Find the intercept on the x axis.

(c) Write the expression for K_c for the reaction

$$Cu(s) + 2Ag^{+}(aq) \rightleftharpoons Cu^{2+}(aq) + 2Ag(s)$$

and use the results for the ionic concentrations at equilibrium to calculate its value. In what units is it expressed?

(d) In Book 1, it is shown that

$$\Delta G^{\ominus}_{298} = -5.7 \log K_c \text{ kJ mol}^{-1}$$

Calculate $\Delta G^{\ominus}$ from your results.

(e) $\Delta G^{\ominus}$ was calculated in Experiment 2.2 from the relation

$$\Delta G^{\ominus}_{298} = -E^{\ominus} \times 2 \times 96.5 \text{ kJ mol}^{-1}$$

How do your results agree?

In this experiment we have assumed that the temperature is approximately 298 K.

Activities

It would have been more correct to have used activities instead of molar concentrations in the above calculations. Because of ionic interactions the activity of an ion in an electrolyte solution is not the same as its molar concentration. It is necessary to multiply the molar concentrations by the appropriate activity coefficient to obtain the activity of the ion. A table of activity coefficients is given in the Appendix (page 63).

Why did we study the effect of changing the concentration of $Ag^{+}(aq)$ rather than $Cu^{2+}(aq)$ in the above cell?

3: Standard Electrode Potentials

3.1 Measuring Standard Electrode Potentials for some Metal/Metal Ion Electrodes

We have seen that when a metal is in contact with a solution of its ions an equilibrium of the type

$$M(s) + aq \rightleftharpoons M^{n+}(aq) + ne^-$$

is established. It seems likely that the extent to which this happens will vary with the metal; indeed the experiments we have carried out so far have shown that this is the case. The electrode potential is a quantitative measure of the position of the equilibrium.

It is not possible to measure such potentials absolutely because a cell must have two electrodes and the effect we measure is always the sum of two electrode potentials. In Experiment 2.3 we compared the potentials of a series of silver electrodes with that of copper as a standard; but the copper electrode is not the internationally accepted standard; this is the standard hydrogen electrode.

In the hydrogen electrode, the electrode reaction involved is

$$H_2(g) + aq \longrightarrow 2H^+(aq) + 2e^-$$

It is obvious that difficulties must arise in attempting to use a gas as an electrode. Fortunately, however, hydrogen is adsorbed by finely divided platinum, and this can be used as a 'support' for the hydrogen. Figure 3.1 shows a form of hydrogen electrode which can be used in the experiments which follow.

The cell notation for a cell of which one electrode is a hydrogen electrode and the other is a metal M in a solution of its ions is

$$Pt[H_2(g)] \,|\, 2H^+(aq) \;\vdots\; M^{n+}(aq) \,|\, M(s)$$

E for this cell is the electrode potential of the metal electrode under the conditions used. When the hydrogen is at one atmosphere pressure and the molar concentrations of $H^+(aq)$ and $M^{n+}(aq)$ are both one, the value of the cell potential is the standard electrode potential of the metal, $E^{\ominus}$. As we have seen, this potential will also vary with temperature, and the temperature taken as standard is 298 K. To be precise, the standard electrode potential of a metal is given the symbol $E^{\ominus}_{298}$, but if the temperature subscript is omitted, standard temperature is assumed. The usual convention regarding the sign of E is, of course, observed.

Why is the potential of the electrode $Pt[H_2(g)] \,|\, 2H^+(aq)$ that of hydrogen and not of platinum?

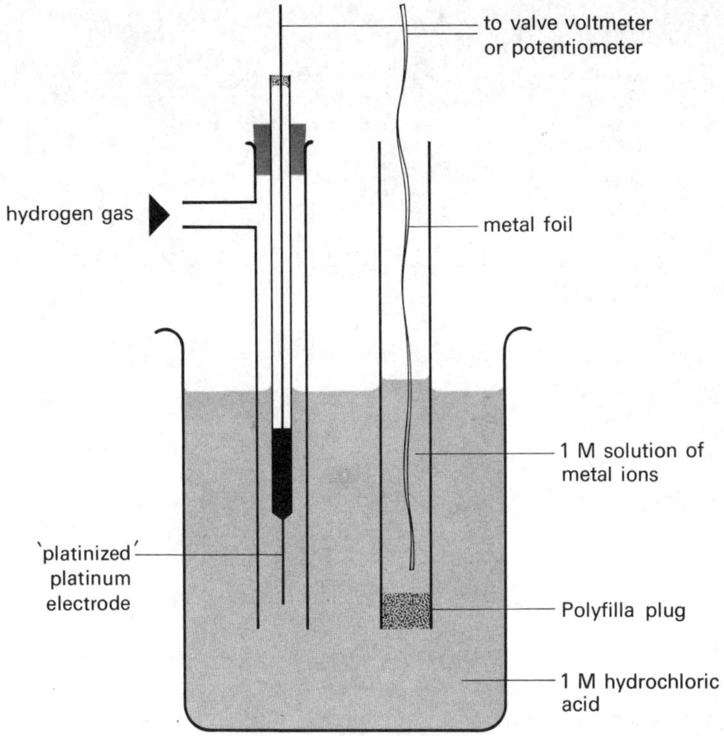

Figure 3.1 Measuring the electrode potentials of some metal/metal ion systems

Apparatus and Materials

Glass tubes (1 cm external diameter × 10 cm)
Polyfilla
Emery cloth
10 per cent solution of potassium nitrate
Valve voltmeter or potentiometer
Connecting leads
Hydrogen electrode
1.00 M solutions of copper(II) sulphate, zinc sulphate, lead nitrate, silver nitrate, hydrochloric acid
Strips of copper, zinc, lead, and silver foil
Supply of hydrogen

Procedure

Set up the apparatus as shown in Figure 3.1. Make the metal/metal ion electrodes as in the previous experiments, cleaning each metal prior to use with emery cloth.

Measure the potential of each cell in turn.

Results

Record your results in a table as follows.

Metal	*Cell notation*	$E^{\ominus}$ *in volts*
Copper		
Zinc		
Lead		
Silver		

Discussion

Compare the values you have obtained with those given in the table on page 61–2. Account for any discrepancies.

Redox Systems

The half cell reactions which take place in a cell, such as

$$H_2(g) - 2e^- \longrightarrow 2H^+(aq) \quad (1)$$

and

$$Cu^{2+}(aq) + 2e^- \longrightarrow Cu(s) \quad (2)$$

in the cell

$$Pt[H_2(g)] | 2H^+(aq) \vdots Cu^{2+}(aq) | Cu(s)$$

are redox reactions. In reaction (1) hydrogen gas is oxidized to hydrogen ions by removal of electrons and in reaction (2) copper ions are reduced to copper by the addition of electrons.

All cell processes are of this type and are therefore redox systems.

3.2 Measuring the Standard Electrode Potentials of Some Other Redox Systems

So far we have limited our work to electrodes involving a metal (reduced form) in contact with a solution of its ions (oxidized form). We are now going to extend this to find the electrode potentials of such systems as

$$Fe^{2+}(aq) - e^- \longrightarrow Fe^{3+}(aq) \quad (1)$$

and

$$2I^-(aq) - 2e^- \longrightarrow I_2(aq) \quad (2)$$

To do this the inert metal platinum is inserted into a mixture of the species concerned, and the e.m.f. of the electrode is determined against a standard hydrogen electrode.

We shall use reactions (1) and (2) at the foot of page 16, and

$$MnO_4^-(aq) + 8H^+(aq) + 5e^- \longrightarrow Mn^{2+}(aq) + 4H_2O(l) \qquad (3)$$

As in our previous work, the potential will vary with concentration, and to obtain standard electrode potentials the concentrations of all species should be 1.00 M. However potassium permanganate is not sufficiently soluble to make a 1.00 M solution, and in this case we shall use a 0.10 M solution.

Apparatus and Materials

Glass tubes (1 cm external diameter × 10 cm)
Polyfilla
Hydrogen electrode and supply of hydrogen
3 platinum electrodes
10 per cent solution of potassium nitrate
Valve voltmeter or potentiometer
1.00 M solution of hydrochloric acid
A solution which is 1.00 M with respect to both iron(II) sulphate and iron(III) sulphate
A solution that is 1.00 M with respect to both iodine and potassium iodide
A solution that is 0.10 M with respect to potassium permanganate, sulphuris acid, and manganese(II) sulphate

Procedure

Set up the apparatus shown in Figure 3.2. Measure the potential of each cell in turn.

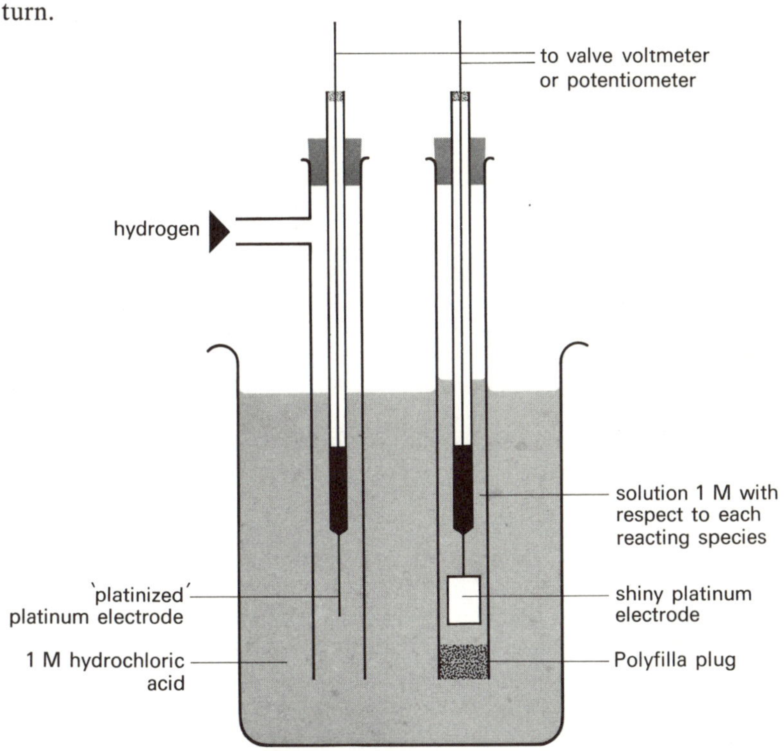

Figure 3.2 Measuring the electrode potentials of some other redox systems

Results

Record your results in a table of the form shown below.

Electrode system	$E^{\ominus}$ *in volts*

Extending the Cell Notation to cover these Systems

Stated generally the cell notation we have used for metal electrodes is

$$\text{reduced form} \mid \text{oxidized form} \; \vdots \; \text{oxidized form} \mid \text{reduced form}$$

Exactly the same notation can be used for the cells made by combining the redox systems studied above. Thus, for the cell shown in Figure 3.3 the notation would be

$$Pt \mid Fe^{2+}(aq),\ Fe^{3+}(aq) \; \vdots \; I_2(aq),\ 2I^-(aq) \mid Pt$$

As before, the sign of E for the cell is conventionally taken as the polarity of the right-hand electrode.

Calculation of Cell Potentials from $E^{\ominus}$ Values for Half-cell Reactions

To calculate the potentials of cells made by combining electrodes such as these you have been investigating in the above experiment, reverse the sign of the potential of the left-hand electrode and add this to the potential of the right-hand electrode.

Using the values you obtained in your experiment, what potential would you expect the cell shown in Figure 3.3 to have?

The three electrode systems you have studied can be combined to make a total of three different cells. Predict the cell potential you would expect for each and enter your results in a table such as the one below. Set up each cell in turn and see whether its experimentally determined potential agrees with your prediction.

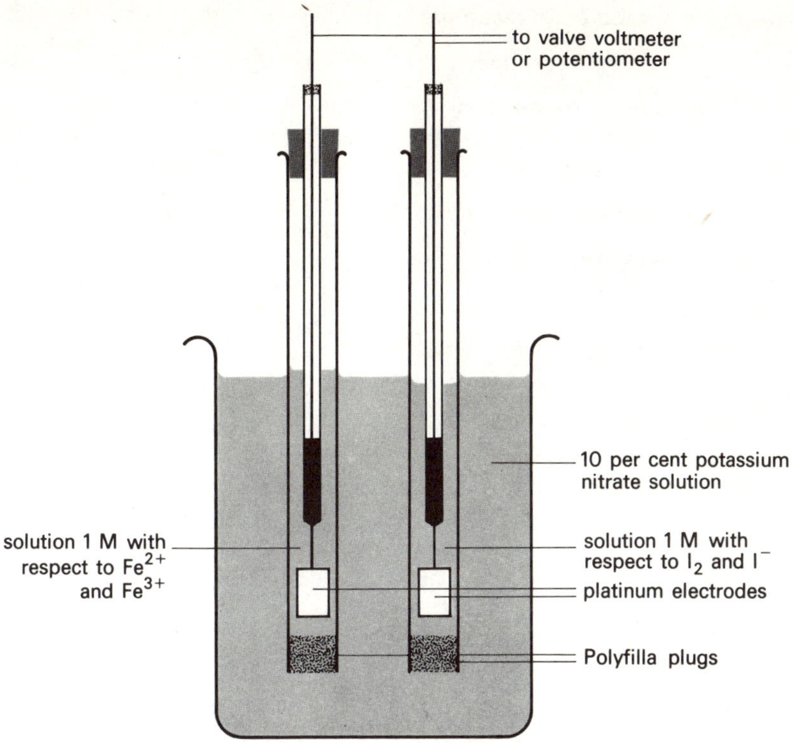

Figure 3.3 Making cells by combining electrodes

Cell notation	*$E^\ominus$ in volts*	
	predicted	*measured*

Use of $E^\ominus$ Values to predict the Feasibility of Redox Reactions

Consider the electrode systems

$$Mg^{2+}(aq)\,|\,Mg(s) \qquad E^\ominus = -2.40 \text{ volts}$$

and

$$Pb^{2+}(aq)\,|\,Pb(s) \qquad E^\ominus = -0.13 \text{ volts}$$

If these are combined to make the cell

$$\mathrm{Pb(s)}\,|\,\mathrm{Pb^{2+}(aq)} \,\vdots\, \mathrm{Mg^{2+}(aq)}\,|\,\mathrm{Mg(s)}$$

the cell will have a potential of -2.27 volts. As the magnesium electrode is the more negative, electrons will be lost from it (oxidation takes place here) and transferred to the lead electrode (where reduction takes place).

Reaction at magnesium electrode: $\mathrm{Mg(s)} - 2e^{-} \longrightarrow \mathrm{Mg^{2+}(aq)}$

Reaction at lead electrode: $\mathrm{Pb^{2+}(aq)} + 2e^{-} \longrightarrow \mathrm{Pb(s)}$

Overall reaction: $\mathrm{Mg(s)} + \mathrm{Pb^{2+}(aq)} \longrightarrow \mathrm{Mg^{2+}(aq)} + \mathrm{Pb(s)}$

To avoid having to use this rather cumbersome reasoning routinely, two rules can be applied which lead to the same result.

(1) *The anti-clockwise rule.* The electrode potentials for the two electrodes are written with the most negative above the other. Anti-clockwise arrows are then written as illustrated below. These indicate the direction of change.

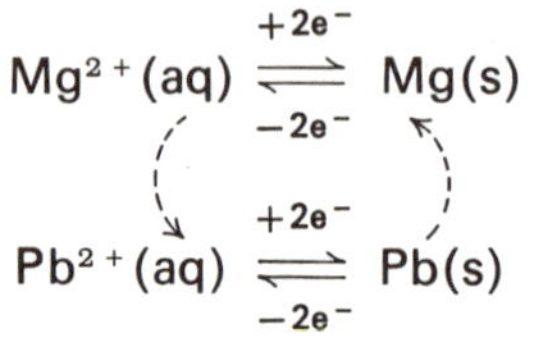

$E^{\ominus} = -2.40$ volts

$E^{\ominus} = -0.13$ volts

(2) *The ABCD rule.* The cell notation is written and the cell potential calculated.

$$\mathrm{A}\,|\,\mathrm{B} \,\vdots\, \mathrm{C}\,|\,\mathrm{D}$$

If E for the cell is *positive* the direction of change is $\mathrm{A} + \mathrm{C} \rightarrow \mathrm{B} + \mathrm{D}$
If E for the cell is *negative* the direction of change is $\mathrm{B} + \mathrm{D} \rightarrow \mathrm{A} + \mathrm{C}$

Check both these rules for accuracy when applied to the magnesium/lead reaction above. The Appendix gives a list of standard redox potentials. Apply these rules to a selection of pairs of half-cell reactions and check your predictions by experiment.

The Nernst Equation

Refer back to the graph you obtained in Experiment 2.3 giving the effect of change of concentration of electrolyte on cell potential. If, instead of using a copper electrode as the standard a hydrogen electrode had been employed, a graph like the one given in Figure 3.4 would have been obtained.

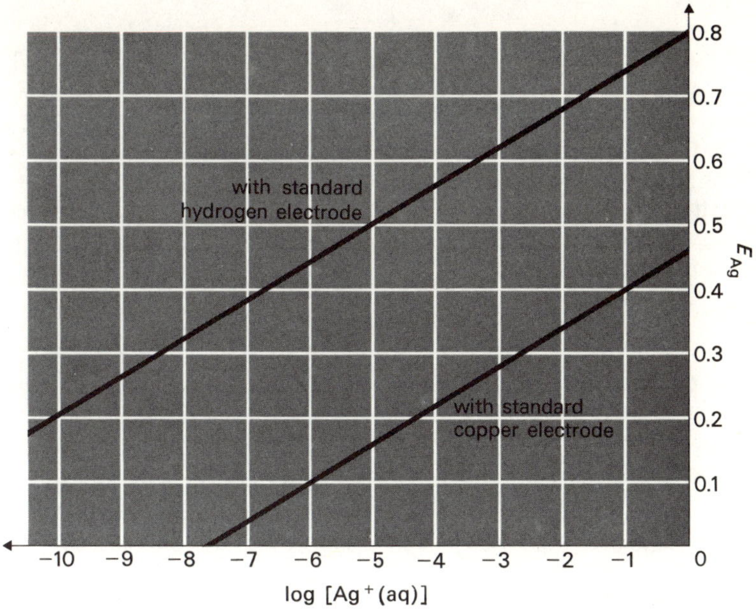

Figure 3.4 Effect on cell potential of concentration changes

Both graphs would have the same slope, but the intercept on the *y* axis would be different (by the standard electrode potential of copper). If we apply the linear law

$$y = c + mx$$

to this, we obtain

$$E_{\text{electrode}} = E^{\ominus}_{\text{electrode}} + 0.06 \log [Ag^{+}(aq)]$$

In the general case, the equation is

$$E_{\text{electrode}} = E^{\ominus}_{\text{electrode}} + \frac{0.06}{z} \log [X^{z+}(aq)]$$

and this is one form of the *Nernst equation for a single electrode*. In your theory course you will have learnt that the more general form of this equation is

$$E_{\text{electrode}} = E^{\ominus}_{\text{electrode}} + \frac{RT}{zF} \ln \frac{[\text{oxidized form}]}{[\text{reduced form}]}$$

In terms of logarithms to the base 10 this is

$$E_{\text{electrode}} = E^{\ominus}_{\text{electrode}} + \frac{2.3RT}{zF} \log_{10} \frac{[\text{oxidized form}]}{[\text{reduced form}]}$$

R, the gas constant, is 8.3 J K^{-1}; T is the temperature in kelvin (298 under standard conditions); z is the number of charges on the ion; F is the faraday = 96 500 coulomb.

Substitute these values in the expression $2.3RT/zF$ and check that this agrees with the slope of the curve you obtained in Experiment 2.3 in both magnitude and dimension.

4: Equilibrium Constants

4.1 Finding the Solubility Products of some Silver Salts using Electrode Potentials

If we know how the e.m.f. of a cell depends on the concentration of ions present, it is obviously possible to find this concentration from determinations of cell potentials. This method is particularly valuable when the concentration of ions is small, and is therefore very difficult to determine by volumetric or gravimetric analysis.

To all intents and purposes silver chloride would be said to be insoluble in water. No substance, however, is completely insoluble, and silver chloride in the presence of water is in equilibrium with Ag^+ and Cl^- ions.

$$AgCl(s) \rightleftharpoons Ag^+(aq) + Cl^-(aq) \qquad (1)$$

The equilibrium lies very far to the left, and the concentration of $Ag^+(aq)$ in equilibrium with AgCl(s) is very small. It can, however, be determined by means of a cell.

The product $[Ag^+][Cl^-]$ is known as the solubility product of silver chloride. It is really the equilibrium constant of the system (1) above, but as AgCl is solid its concentration is invariable, and does not enter into the equilibrium equation.

Apparatus and Materials

- 5 cm^3 and 10 cm^3 pipettes
- Access to centrifuge and centrifuge tubes
- Glass tubes (1 cm external diameter × 10 cm)
- Strip of copper and strip of silver foil
- Valve voltmeter or potentiometer
- Connecting leads
- Polyfilla
- 10 per cent solution of potassium nitrate
- 0.10 M solutions of copper(II) sulphate, silver nitrate, potassium chloride, potassium bromide, potassium iodide, and potassium iodate

Procedure

Place 5.0 cm^3 of 0.10 M silver nitrate and 10.0 cm^3 of 0.10 M potassium chloride in a centrifuge tube, mix well, and separate the precipitate. Pour the supernatant liquid carefully into the silver electrode compartment of the cell shown in Figure 4.1, and determine its potential.

Use the graph in Figure 3.4 as a calibration curve to determine the silver ion concentration. This is $[Ag^+(aq)]_{eqm}$.

Calculation

As you will see, there has been *virtually* complete precipitation of the silver ions. This would use up half of the chloride ions originally present. The

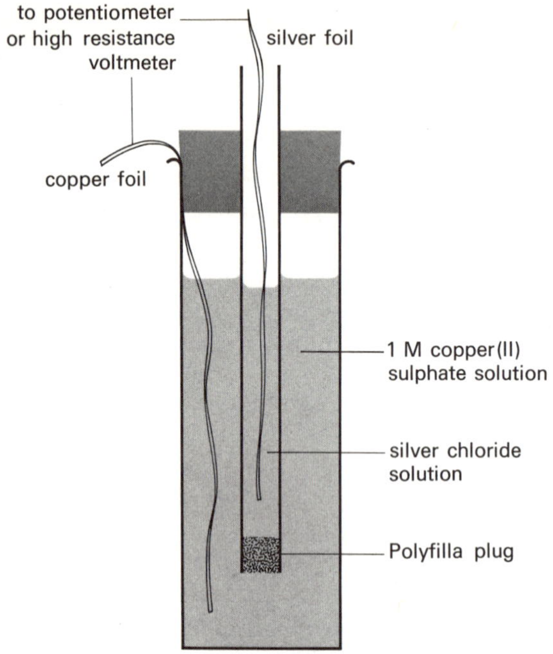

Figure 4.1 Determining the solubility product of some silver salts

remaining chloride ions have been diluted by two-thirds because of the addition of 5 cm^3 of silver nitrate to 10 cm^3 of the original potassium chloride. Hence

$$[Cl^-(aq)]_{eqm} = 0.1 \times 0.5 \times 0.67 = 0.033 \text{ mol dm}^{-3}$$

Use these results to calculate the solubility product of silver chloride:

$$K_{s.p.} = [Ag^+(aq)]_{eqm}[Cl^-(aq)]_{eqm}$$

Repeat the experiment using 5.0 cm^3 of 0.10 M silver nitrate each time with 10.0 cm^3 of 0.10 M potassium bromide, potassium iodide, and potassium iodate respectively.

Questions

1. What approximation has been made in the calculation? Does it seriously affect the values obtained for the solubility products?
2. Do the results agree with your previous knowledge of the relative solubilities of silver chloride, bromide, iodide, and iodate?
3. Show in outline, without doing the actual calculation, how it would be possible to obtain the solubility of a salt such as silver chloride from its solubility product.
4. What cell could you set up to determine the solubility product of lead(II) chloride?

4.2 Finding the Equilibrium Constant for a Redox Reaction

In this experiment we again use a cell to determine the concentration of an ion. The reaction for which the equilibrium constant is to be determined is the oxidation of iron(II) ions by silver ions.

$$Ag^{+}(aq) + Fe^{2+}(aq) \rightleftharpoons Ag(s) + Fe^{3+}(aq)$$

There are several ways of carrying out this determination. We can find the concentration of Ag^{+}(aq) at equilibrium by an electrochemical method, and by volumetric analysis by titration against standard potassium thiocyanate solution. We shall use these two methods in this experiment.

Apparatus and Materials

Glass tubes (1 cm external diameter × 10 cm)
Platinum electrode
Strip of silver foil
Valve voltmeter or potentiometer
100 cm³ conical flasks
25 cm³ pipette
150 cm³ beaker with cork bored to take electrodes (see Figure 4.2)
50 cm³ burette
Pipettes for making dilutions
0.10 M iron(III) nitrate made up in 0.05 M nitric acid
0.20 M iron(II) sulphate solution
0.20 M barium nitrate solution
0.40 M silver nitrate solution
0.10 M silver nitrate solution
0.10 M potassium thiocyanate solution
Polyfilla
10 per cent potassium nitrate solution

Procedure

(1) *Electrochemical determination*

Put 25.0 cm³ of 0.20 M iron(II) sulphate and 25.0 cm³ of 0.20 M barium nitrate in a beaker. Barium sulphate will be precipitated leaving 50.0 cm³ of 0.10 M iron(II) nitrate solution. Add to this 50.0 cm³ of the 0.10 M iron(III) nitrate solution. When the precipitate has settled the supernatant liquid is 0.05 M with respect to both iron(II) and iron(III) nitrates. Make up silver electrodes which have silver ion concentrations of 0.40, 0.20, 0.10, 0.05, and 0.025 M by successive dilution. Assemble each in turn in the cell illustrated in Figure 4.2 and determine its potential.

(2) *Volumetric determination*

Place 50.0 cm³ of 0.20 M iron(II) sulphate solution and 50.0 cm³ of 0.20 M barium nitrate solution into a conical flask. Mix well, cork, and allow the precipitate to settle. The supernatant liquid is a 0.10 M solution of iron(II) nitrate. Pipette 25.0 cm³ of this solution into each of three conical flasks and add to each 25.0 cm³ of 0.10 M silver nitrate solution. Cork each flask and allow to stand overnight in a dark cupboard in order to allow the reaction to attain equilibrium.

Without disturbing the precipitate of silver, pipette 25.0 cm³ from one flask into a fresh conical flask and carry out a pilot titration to find the

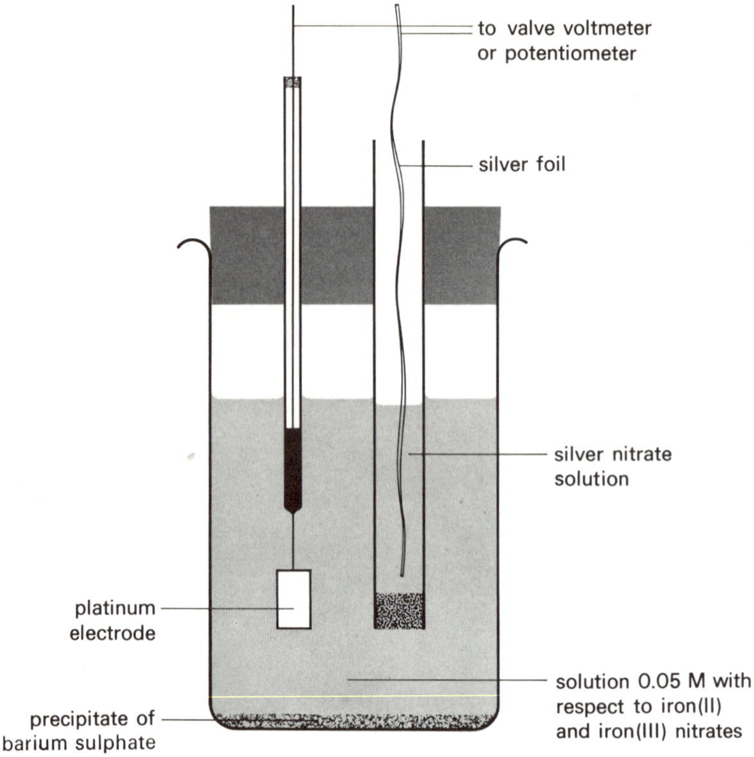

Figure 4.2 Electrochemical determination of an equilibrium constant

volume of 0.10 M potassium thiocyanate solution that must be added before a blood-red coloration is formed in the reaction mixture. The silver ions in the equilibrium mixture react with the thiocyanate ions according to the equation

$$Ag^+(aq) + CNS^-(aq) \longrightarrow AgCNS(s)$$

When virtually all the silver ions have been removed, the thiocyanate ions in excess react with the iron(III) ions also present in the equilibrium mixture to give the blood-red colour of the $FeCNS^{2+}(aq)$ complex that indicates the end point:

$$Fe^{3+}(aq) + CNS^-(aq) \longrightarrow \underset{\text{(blood-red)}}{FeCNS^{2+}(aq)}$$

Use the remaining two flasks of equilibrium mixture to determine the accurate titre.

Results

(1) *Electrochemical determination*
Enter your results in a table such as that shown below, making sure that the sign given to E is in accordance with the cell notation

$$\mathrm{Pt}|\mathrm{Fe^{2+}(aq), Fe^{3+}(aq)} \vdots \mathrm{Ag^{+}(aq)}|\mathrm{Ag(s)}$$

$[Ag^{+}(aq)]$ *in mol* dm^{-3}	*log* $[Ag^{+}(aq)]$	E_{cell} *in volts*

Plot a graph of E_{cell} (y axis) against log $[Ag^{+}(aq)]$ (x axis). From the graph find the equilibrium concentration of silver ion, $[Ag^{+}(aq)]_{eqm}$, and use this to determine K_c for the reaction.

$$K_c = \frac{[Fe^{3+}(aq)]_{eqm}}{[Ag^{+}(aq)]_{eqm}[Fe^{2+}(aq)]_{eqm}}$$

When

$$\frac{[Fe^{3+}(aq)]_{eqm}}{[Fe^{2+}(aq)]_{eqm}} = 1, \text{ (as in this experiment)}, \quad K_c = \frac{1}{[Ag^{+}(aq)]_{eqm}}$$

Calculation

(2) *Volumetric determination*
(a) Calculate the molarity of the silver ions in the reaction mixture. As one silver ion reacts with one iron(II) ion, this is also the molarity of iron(II) ions remaining in the mixture.

(b) If complete reaction had taken place the molarity of iron(III) ions would have been 0.05—remember that an equal volume of silver nitrate solution was added to the original 0.10 M solution. Calculate the actual iron(III) ion concentration, allowing for that which we know to be unchanged as iron(II).

(c) Calculate K_c for the reaction

$$K_c = \frac{[Fe^{3+}(aq)]_{eqm}}{[Fe^{2+}(aq)]_{eqm}[Ag^{+}(aq)]_{eqm}}$$

Questions

1. Examine the statements made about the concentrations of solutions after mixing in both the electrochemical and volumetric methods. Are these correct? If not, what effect are the assumptions likely to have on the results?

2. Compare the results of the two methods. Are they in satisfactory agreement? Which do you consider to be the more accurate method?
3. What other method could be used to investigate this equilibrium? Suggest other equilibria that could be investigated electrochemically and devise the necessary cell set-ups.

5: Uses of pH Measurements

5.1 Electrometric Measurement of pH and the Use of pH Measurements to obtain some Acid–Base Equilibrium Constants

Consider the possible experimental set-up illustrated in Figure 5.1. The electrode potential for electrode A is, by definition, zero. The electrode potential for electrode B is given by the Nernst equation

$$E = E^{\ominus} + 0.06 \log [H^+(aq)]$$

Now, pH = $-\log [H^+]$ and hence $E = -0.06$ pH. Such an assembly could therefore be used for determining pH.

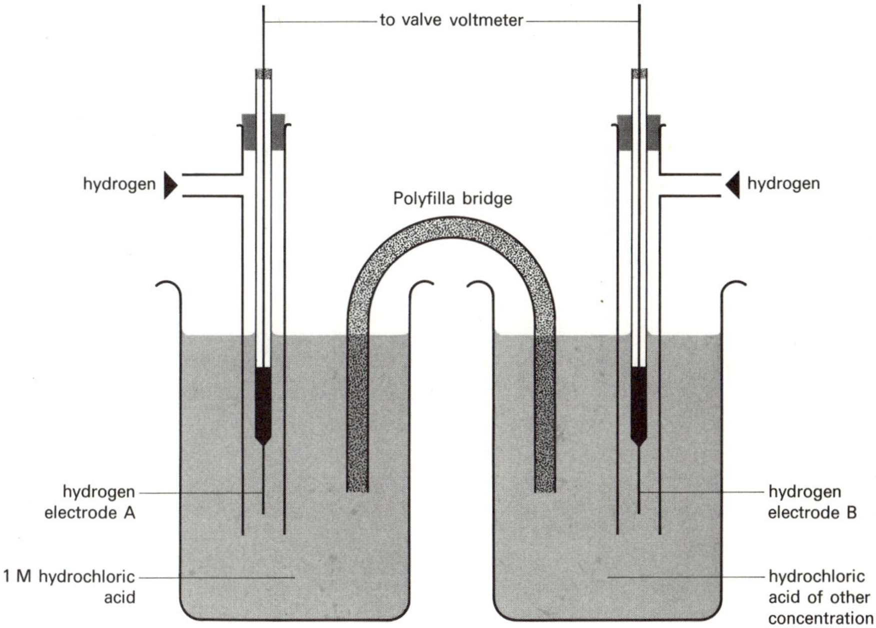

Figure 5.1

pH meters

This cell is, however, cumbersome to set up, difficult to transport, and requires a considerable volume of test liquid. Modern pH meters have alternative electrode systems that overcome these problems. The electrode A instead of being a hydrogen electrode is replaced by another reference electrode. A length of silver wire coated with silver chloride and dipping into a saturated solution of potassium chloride is widely used for this purpose. It develops a very steady electrode potential which, however, is not zero. Hence calibration against a solution of known pH is essential.

The test electrode, B, is replaced by a glass electrode. The theory of the glass electrode is complex, and advanced texts often contain conflicting accounts of how it works. Figure 5.2 shows such an electrode system assembled to determine the unknown pH of a liquid.

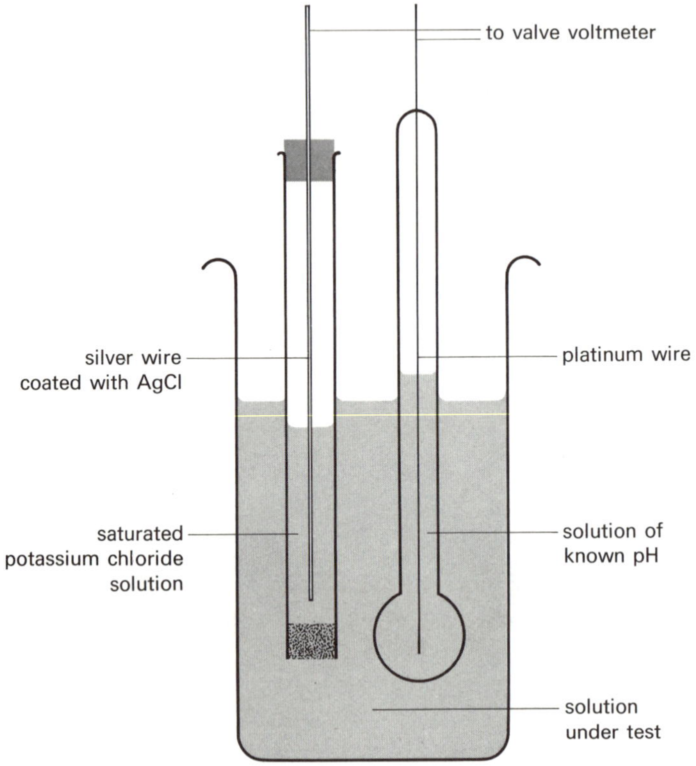

Figure 5.2 Glass and reference electrodes to measure pH

Examine the electrodes of the pH meter available in your school. They are delicate and must be handled with care. Do NOT touch the bulb of the glass electrode. You may find that in some instruments the two electrodes are combined in one unit, the reference electrode being assembled around the glass electrode. As the e.m.f. falls drastically when current is taken from cells using such electrodes, they cannot be used with a potentiometer. A valve voltmeter must be used.

The detailed operation of individual pH meters varies from model to model and so the manufacturer's instructions must be consulted concerning calibration and operation. Remember that when the electrodes are transferred from one liquid to another they must be carefully washed with distilled water before being placed in the second liquid.

You will now use a pH meter to determine the equilibrium constant for the ionic dissociation of ethanoic acid (acetic acid) according to the equation

$$CH_3CO_2H(aq) \rightleftharpoons CH_3CO_2^-(aq) + H^+(aq)$$

and the hydrolysis constant for the action of water on ethanoate ions (acetate ions) according to the equation

$$CH_3CO_2^-(aq) + H_2O(l) \rightleftharpoons CH_3CO_2H(aq) + OH^-(aq)$$

Apparatus and Materials

pH meter and electrodes correctly adjusted

25 cm^3 pipette

1.00 M solutions of ethanoic acid and sodium ethanoate

Procedure

Determine the pH of the two solutions as accurately as you can with the pH meter.

Calculation

(a) Refer back to the equation for the ionic dissociation of ethanoic acid given above. The equilibrium constant of this reaction, K_c will be

$$K_c = \frac{[CH_3CO_2^-(aq)]_{eqm}[H^+(aq)]_{eqm}}{[CH_3CO_2H(aq)]_{eqm}}$$

This is frequently known as the dissociation constant of the acid and is given the symbol K_a. Calculate $[H^+(aq)]_{eqm}$ from the pH for 1.00 M ethanoic acid. What will be the concentration of ethanoate ions?

If we assume that these very small concentrations have not reduced the concentration of undissociated ethanoic acid we can write $[CH_3CO_2H(aq)] = 1.00$. Use these figures to calculate a value for K_c.

(b) The hydrolysis constant for the reaction of ethanoate ions in a solution of sodium ethanoate with water is given by

$$K_h = \frac{[CH_3CO_2H(aq)]_{eqm}[OH^-(aq)]_{eqm}}{[CH_3CO_2^-(aq)]_{eqm}}$$

From the measured value of the pH of 1.00 M sodium ethanoate calculate the hydroxide ion concentration, $[OH^-(aq)]_{eqm}$. To do this you will need to use the ionic product for water

$$K_w = [H^+(aq)]_{eqm}[OH^-(aq)]_{eqm} = 10^{-14}$$

Use these figures to calculate a value for K_h.

Discussion and further practical work

You can use your numerical value for K_c to calculate the pH of any solution that is equimolar with respect to both ethanoic acid and ethanoate ions. Do this, and then mix 25.0 cm^3 each of 1.00 M ethanoic acid and sodium ethanoate and find the pH. How does it compare with the value you calculated? The strength of an acid is given by the hydrogen ion concentration of its aqueous solution. By finding the pH of solutions of different acids of the same concentration you can place them in order of strength. Do this with the series of acids, methanoic (formic), ethanoic, propanoic (propionic), and butanoic (butyric), and with the series chloroethanoic, dichloroethanoic, and trichloroethanoic.

5.2 Investigating the pH Changes during the Course of Acid–Base Titrations

When an acid is added to a base the reaction

$$H^+(aq) + OH^-(aq) \rightleftharpoons H_2O(l)$$

takes place. The equilibrium lies far to the right. Hence as alkali is added to acid there is a progressive diminution of the hydrogen ion concentration, and it should be possible to follow this by means of a pH meter. The experiment is best carried out on a class co-operative basis; each group of students should deal with one of the acid–base combinations.

Apparatus and Materials

25 cm^3 pipette
50 cm^3 burette
Magnetic stirrer and 'slave'
pH meter and electrodes

1.00 M solutions of ammonia, sodium hydroxide, hydrochloric acid, sulphuric acid and ethanoic acid

Procedure

Place 25.0 cm^3 of the acid in a 100 cm^3 beaker. Stand this on the magnetic stirrer and put the 'slave' in position. Clamp the pH electrodes so that they are in the solution. Fill the burette with alkali (remembering to see that the nozzle of the burette is full), and clamp it so that the alkali can be discharged into the acid. Note the pH of the solution.

Switch on the magnetic stirrer and run in 1.0 cm^3 portions of alkali, recording the pH after each addition. A total of 30 cm^3 of alkali should be added in this way, except in the case of sulphuric acid when 60 cm^3 of alkali should be added in 2.00 cm^3 portions.

Is there a time during the course of the addition when very little alkali seems to have a great effect on the pH? If so, repeat the experiment but this time add the alkali *a drop at a time* over this critical stage.

Results

Pool the class results and plot them using axes as shown in Figure 5.3. Use a different coloured pencil for each combination of acid and alkali.

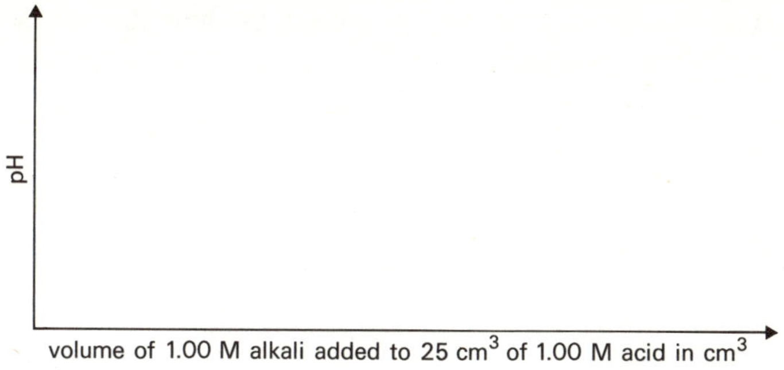

Figure 5.3

Alkaline solutions attack glass. At the end of the experiment rinse out the burette with bench dilute sulphuric acid and then with distilled water.

Discussion

Explain the types of curve obtained. Why is the curve for sulphuric acid so different from the others?

5.3 Investigating the Working Ranges of some pH Indicators

Indicators change colour at different pH values. In this experiment you will find out the pH ranges for which a number of common indicators are useful. For this you will need solutions of known pH and of which the pH is unchanged when small quantities of other substances, such as the indicators, are added. Such solutions are called 'buffer' solutions.

Apparatus and Materials

Rack of 8 test-tubes
Buffer solutions of pH 3, 4, 5, 6, 7, 8, 9 and 10
A number of acid–base indicators chosen from the following: bromophenol blue, bromocresol green, methyl red, chlorophenol red, bromocresol purple, bromothymol blue, phenol red, methyl orange, diphenol purple, phenolphthalein, cresol red, *p*-nitrophenol, and phenol violet
Universal indicator
A supply of coloured crayons would be useful

Procedure

Half fill the test-tubes with the range of buffer solutions one to each tube, and arrange them in ascending value of pH. To each buffer solution add two drops of the first indicator you decide to test. Note the colour of the indicator at each pH. Repeat the experiment with as many more indicators as you can find time for. Include among them the universal indicator. Make a chart, as shown in Figure 5.4, preferably in colour, to show the pH range over which each indicator changes colour.

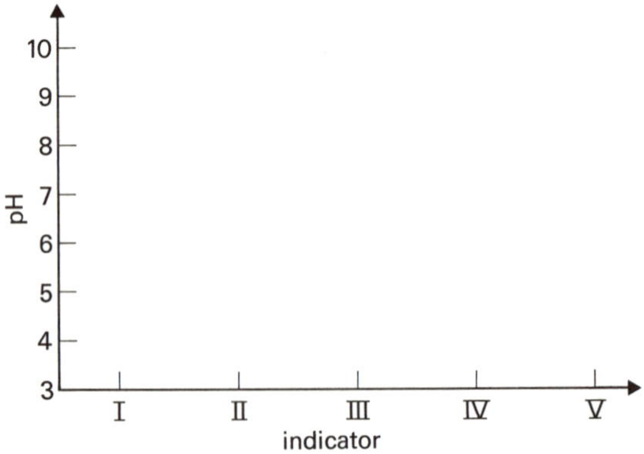

Figure 5.4

Questions

1. Using your results from Experiment 5.2 which indicators would be useful for the various neutralizations you studied?
2. Do the colour changes of the universal indicator correspond with those shown on the chart supplied with the indicator?
3. Can you make up a universal indicator from the results for individual ones? Try to do this, and test it out with the buffer solutions.

6: A Simple Fuel Cell

6.1 Investigating a Simple Fuel Cell

A fuel cell is an electrochemical device which utilizes the energy evolved when a substance normally used as a fuel is oxidized. In the particular case under study hydrogen gas is oxidized by oxygen gas. You will know that this is a reaction which gives out a great deal of heat. Some of this energy (about 80 per cent of it) can be converted directly into electrical energy.

$$2H_2(g) + O_2(g) \longrightarrow 2H_2O(l) \qquad \Delta H^{\ominus} = -572 \text{ kJ mol}^{-1}$$

$$\Delta G^{\ominus} = -474 \text{ kJ mol}^{-1}$$

The cell is composed of two carbon rods, one of which is surrounded with hydrogen and the other with oxygen, immersed in a 5 M solution of potassium hydroxide. The reaction takes place through the intermediate formation and reduction of hydroxyl ions. The electrode reactions are:

$$\text{Oxygen electrode: } O_2(g) + 2H_2O(l) + 4e^- \longrightarrow 4OH^-(aq)$$

$$\text{Hydrogen electrode: } 2H_2(g) + 4OH^-(aq) \longrightarrow 4H_2O(l) + 4e^-$$

On addition these give the overall equation stated above.

Apparatus and Materials

2 carbon electrodes constructed as shown in Figure 6.1
500 cm^3 beaker
Cylinders of oxygen and hydrogen, with control gear
Valve voltmeter
20 kΩ rheostat
Connecting leads and switch
0–1 milliammeter
High impedance input transistor radio (if available)
5 **M** (approximately) solution of potassium hydroxide

Procedure

Warning. 5 M potassium hydroxide is strongly caustic.

Set up the cell and circuit as shown in Figures 6.1 and 6.2. With the switch open determine the voltage of the cell with the valve voltmeter. Under these conditions the cell is providing no current. Now close the switch and find the cell potential when it is supplying various currents.

If available, connect a high impedance input transistor radio (with the correct polarity) to the fuel cell and see if it operates. The usual working voltage for these is 9 volts and the sound may therefore be a little thin!

Discussion

From the $\Delta G^{\ominus}$ value for the reaction given above, calculate the expected e.m.f. of the cell using the relationship $\Delta G^{\ominus} = -zFE^{\ominus}$. How does it compare

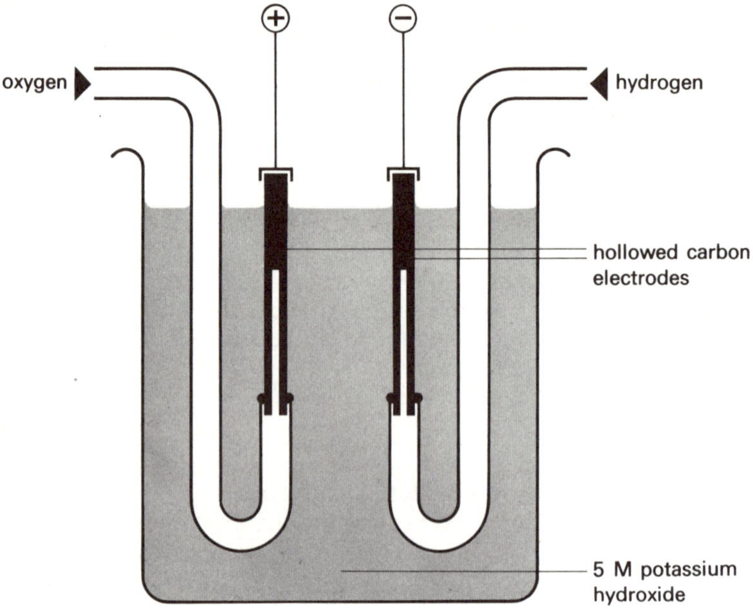

Figure 6.1 A fuel cell

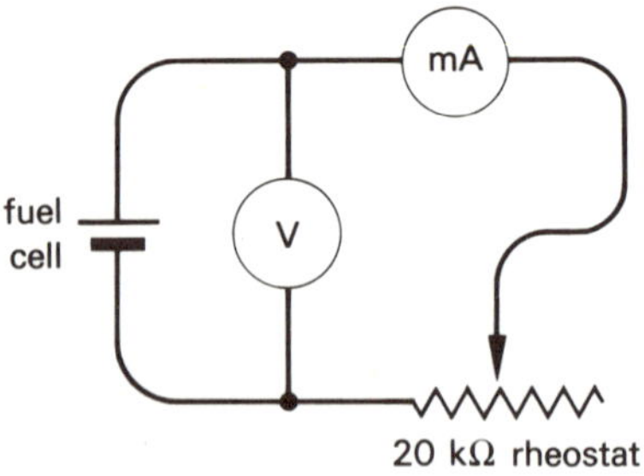

Figure 6.2 Testing a fuel cell

with the value you obtained experimentally? Remembering that the usual way of utilizing the energy obtainable from the combustion of a fuel and converting it into electrical energy is to use the heat to raise steam and then use the steam to turn generators, discuss the advantages of the fuel cell.

7: Electrolysis

7.1 What Products can be obtained by the Electrolysis of Sodium Chloride under Various Conditions?

The electrolysis of sodium chloride is a very important industrial process. In this series of experiments you will investigate the electrolysis of sodium chloride, first when molten, and then in solution, and will find out something of the variety of products that can be obtained from this process.

(a) The electrolysis of molten sodium chloride

Sodium chloride melts at 808 °C. The melting point can be lowered to about 600 °C by adding about 10 per cent by mass of strontium chloride (which actually has a melting point of 873 °C!). This temperature can be attained by using a very good Meker burner, but it is necessary to surround the crucible and burner with fire bricks as otherwise currents of air may reduce the temperature.

Apparatus and Materials

Safety goggles (these must be worn throughout the experiment)
Silica crucible
Carbon electrode
Steel electrode fitted with glass tube, as shown in Figure 7.1
Crucible furnace (i.e. Meker burner and arrangement of fire bricks)
12 volt d.c. supply
Universal indicator (pH) paper
Connecting leads
Sodium chloride
Strontium chloride

Procedure

Set up the apparatus as shown in Figure 7.1. Heat until the mixture of sodium chloride and strontium chloride is molten. Pass the current for some time. While continuing heating, carefully raise the cathode. Some salt will solidify over the open end and seal the sodium formed from the air. When the electrode and glass tube have cooled to room temperature, wrap them in a piece of stout paper and carefully break the glass. Add the pieces to water in a small beaker and observe the evolution of gas. What is the effect of the solution on universal indicator paper?

Questions

1. What are the electrode reactions in this process?
2. Find out from books of data the values of $\Delta G^{\ominus}$ and $\Delta H^{\ominus}$ for the reaction.

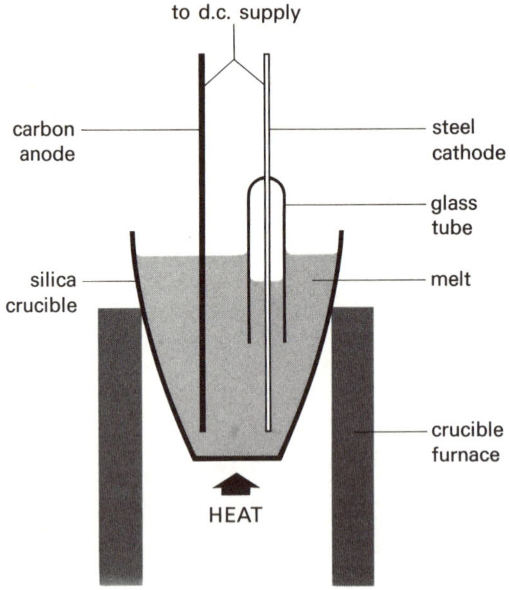

Figure 7.1 Electrolysis of molten sodium chloride

3. What e.m.f. is theoretically required to carry out this electrolysis?
4. Why is $\Delta G^\ominus$ not the sum of the ionization energies of sodium and chlorine?

(b) The electrolysis of cold, dilute sodium chloride solution, the products being separated from each other

Apparatus and Materials

Porous pot fitted with rubber bung bored to carry a carbon electrode, and an L-shaped glass outlet tube as illustrated in Figure 7.2
Connecting leads
Starch–iodide paper
12 volt d.c. supply
500 cm^3 beaker
250 cm^3 bottle to store product
Copper wire cathode
2 M sodium chloride solution

Procedure

Set up the apparatus as in Figure 7.2. Pass the current until no further chlorine issues from the outlet of the anode compartment (test with starch–iodide paper). Disconnect the electrical supply, discard the contents of the anode compartment, and fill the bottle with the solution in the beaker. Keep the solution for the experiments in part (e).

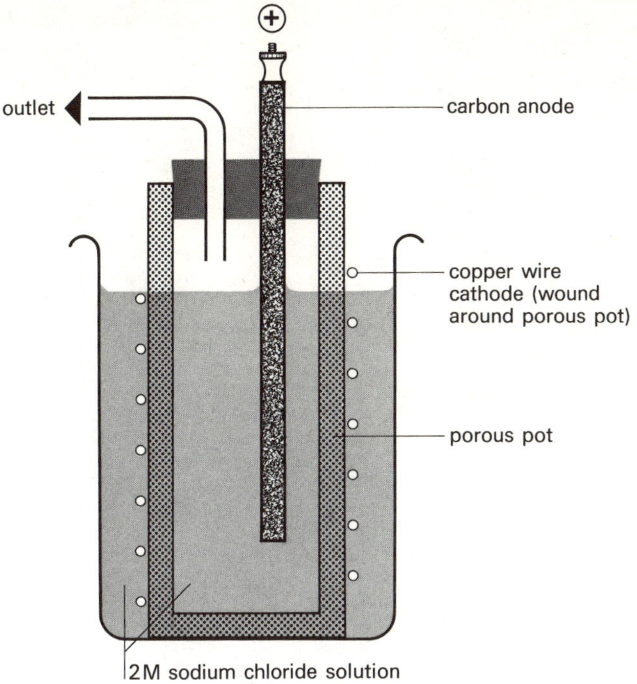

Figure 7.2 Electrolysis of sodium chloride solution, keeping products separate

Discussion

The reactions at the electrodes are

$$\text{At the cathode: } 2H^+(aq) + 2e^- \longrightarrow H_2(g)$$
$$\text{At the anode: } 2Cl^-(aq) - 2e^- \longrightarrow Cl_2(g)$$

There are also present, of course, Na^+ ions from the sodium chloride and OH^- ions from the water. These have not entered into any electrode reaction. Why not?

Is the same e.m.f. required to electrolyse sodium chloride solution as molten sodium chloride? If not, why?

Additional experiment

Try electrolysing a 0.1 M solution of sodium chloride in the above apparatus. Is there any difference in the products? Explain.

(c) The electrolysis of cold, dilute sodium chloride with mixing of the products

In this experiment the products of the reactions at the anode and cathode are allowed to mix in the cold. The electrode reactions are the same as those which occurred in (b), but chlorine is allowed to react with the hydroxyl ions in the bulk of the solution.

Apparatus and Materials

Mini-stirrer
12 volt d.c. supply
500 cm³ beaker
250 cm³ bottle to store product
Connecting leads
Copper cathode
Carbon anode
2 M sodium chloride solution

Procedure

Set up the apparatus as shown in Figure 7.3. Pass the current through the apparatus for twenty minutes maintaining the stirring during the electrolysis. Use the product for the tests in part (e) opposite.

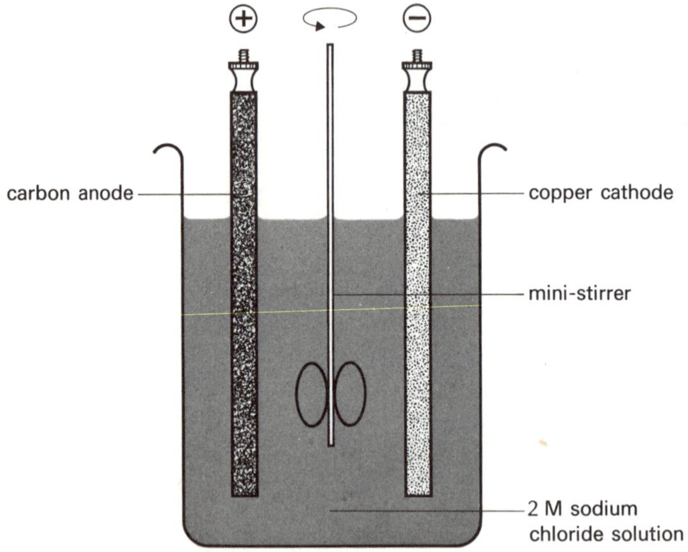

Figure 7.3 Electrolysis of sodium chloride, mixing the products

Discussion

Write the equation for the reaction occurring. Sodium hypochlorite cannot be obtained as a solid and exists only in solution. Solutions made by the electrolysis of brine are sold as household bleaches.

(d) The electrolysis of hot, concentrated sodium chloride solution with mixing of the products

Apparatus and Materials

As for (c) above but substituting 4 M sodium chloride for 2 M
Tripod, gauze, and Bunsen burner

Procedure

Repeat the experiment as described in (c) but keep the temperature of the liquid just below boiling.

Pour a little of the product into an evaporating dish and evaporate carefully to dryness. Put the solid product into a small ignition tube and heat. Test the gas evolved.

Discussion

The hypochlorite ions undergo further reaction under the conditions of this experiment. Write the equation. Your test with the solid product should give you a clue as to what this product is.

In the older textbooks it would have been said that some hypochlorite ions are oxidized and some are reduced in this reaction. Is this really the case?

(e) Tests for the products of electrolysis of sodium chloride solution

We want to show that the product of part (b) contains excess of OH^- ions while that from (c) contains OCl^- ions. We shall therefore try the action of sodium hydroxide and sodium hypochlorite on various solutions of metal ions and then compare the results with those we obtain with the solutions prepared in (b) and (c).

Apparatus and Materials

Rack of test-tubes
Solutions of iron(II) sulphate, cobalt(II) nitrate, manganese(II) chloride, and lead(II) nitrate
Sodium hydroxide solution
Sodium hypochlorite solution
The products of parts (b) and (c)

Procedure

Place about ten drops of each of the solutions above in different test-tubes, and add sodium hydroxide solution to each one a few drops at a time until it is in excess. Record your observations carefully, writing equations wherever possible for the reactions taking place. Repeat the process with sodium hypochlorite solution, and then with the products of electrolysis obtained in parts (b) and (c). What do you conclude?

7.2 How many Coulombs are required to Transfer One Mole of Electrons in Electrolysis?

We shall carry out this determination using the electrolysis of copper(II) sulphate solution. The copper(II) ion bears two positive charges. Hence in the deposition of one mole of copper atoms, two moles of electrons are transferred. The mole of copper has a mass of 63.5 g. So, by weighing the copper deposited when a known quantity of electricity is passed through a solution of copper(II) sulphate, we can obtain by simple proportion the number of coulombs required to transfer one mole of electrons.

Apparatus and Materials

1 litre beaker
2 copper electrodes approximately 10 × 15 cm
3 beakers, large enough to accept the electrodes, containing respectively bench sodium hydroxide, dilute sulphuric acid, and methylated spirits
Emery cloth
Wash-bottle containing distilled water
0–1 ampere ammeter
20 ohm rheostat (high current rating)
Stop-clock
12 volt d.c. supply
Connecting wires
Access to balance reading to 1 mg
Solution of 250 g of copper(II) sulphate in 800 cm^3 of water and 6 cm^3 of concentrated sulphuric acid

Procedure

Each electrode must first be thoroughly cleaned as follows. Remove all surface contaminants with emery cloth and rinse with distilled water. Steep for one minute in sodium hydroxide solution and then rinse with distilled water. Steep for one minute in dilute sulphuric acid and again rinse with distilled water. Rinse with methylated spirits and allow to dry thoroughly.

Connect the electrodes in the circuit shown in Figure 7.4 and pass a current of approximately 0.5 ampere for five minutes. Remove the electrodes, rinse them with distilled water and then with methylated spirits. When they are thoroughly dry weigh them accurately, noting carefully which is to be used as the cathode and which the anode.

Replace the electrodes in the circuit and pass a constant current of 0.75–1 ampere for as long as possible. Record accurately the time for which the current is passed. It is essential that the current should be steady and any variations should be immediately corrected by means of the rheostat. Remove the electrodes, rinse them with distilled water, then with methylated spirits. Dry them and reweigh. Calculate the loss in mass of the anode and the gain in mass of the cathode. If these are not the same which should be regarded as the more reliable? Calculate the number of coulombs (ampere seconds) which have passed through the solution.

Results

Draw up a table of results and calculate the number of coulombs required to transfer 0.5 mole of copper ions.

Questions

1. Write the equations for the reactions occurring at the anode and the cathode, and see that you understand from them why two moles of electrons are associated with the transfer of one mole of copper from anode to cathode. How many moles of electrons are associated with the transfer of one mole of silver from anode to cathode?

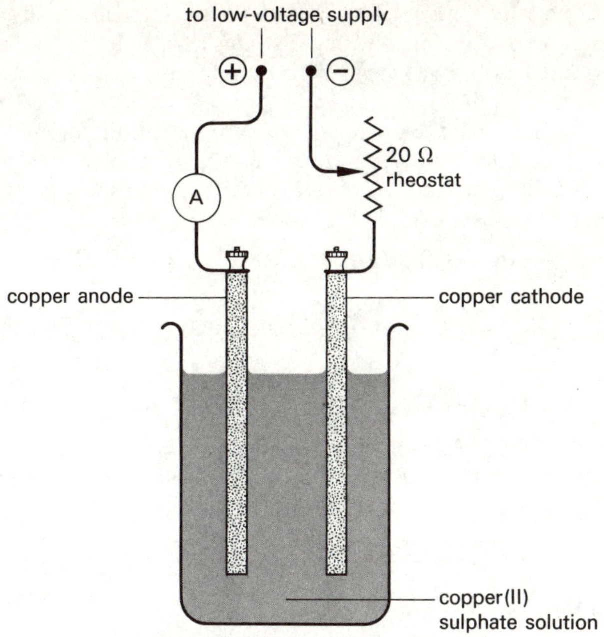

Figure 7.4 Determining a value for the faraday

2. What are the conditions necessary to obtain an accurate result in this experiment? Why was it necessary to be so careful to clean the electrodes?
3. Calculate from your result the charge on the electron.
4. How could experiments of this kind give information about the number of electrons in the outermost shell of a metal atom? Consider carefully whether they give absolute information or only a comparison of the number.
5. This seems a very easy way of finding the charge on the electron. Why, then, was it necessary for Millikan to determine this constant by the much more complicated oil-drop method?
6. The number of coulombs required to transfer one mole of electrons is called the faraday. Why is the experiment best carried out in the way described instead of electrolysing copper(II) sulphate solution between carbon electrodes?

7.3 The Electrolytic Preparation of Sodium Vanadium Alum, $NaV(SO_4)_2{,}12H_2O$

Reactions at anodes and cathodes in electrolysis are respectively oxidations and reductions since at the anode electrons are removed and at the cathode they are added. Hence electrolytic methods are often used to bring about redox reactions which would be difficult to accomplish in any other way. The electrolytic extraction of some metals, such as aluminium and magnesium, which could not be obtained from their ores by using the common reducing agents, are cases in point.

In this experiment we are preparing sodium vanadium alum, $NaV(SO_4)_2$,-$12H_2O$, from sodium metavanadate, $NaVO_3$. This involves a two-stage reduction. The first, to the VO^{2+} stage, is accomplished by conventional chemical reduction, using sulphur dioxide.

$$2VO_3^-(aq) + 4H^+(aq) + SO_2(g) \longrightarrow 2VO^{2+}(aq) + SO_4^{2-}(aq) + 2H_2O(l)$$

What change in the oxidation number of vanadium has taken place here?

The second stage is accomplished electrolytically, and at the cathode the reaction is

$$VO^{2+}(aq) + 2H^+(aq) + e^- \longrightarrow V^{3+}(aq) + H_2O(l)$$

What change in the oxidation number of vanadium has taken place here? The V^{3+} ions are in the presence of $Na^+(aq)$ ions and $SO_4^{2-}(aq)$ ions and crystals of the alum will deposit from the solution.

$$Na^+(aq) + V^{3+}(aq) + 2SO_4^{2-}(aq) + 12H_2O(l) \longrightarrow NaV(SO_4)_2, 12H_2O(s)$$

Apparatus and Materials

250 cm^3 beaker
50 cm^3 measuring cylinder
Drechsel wash-bottle
Porous pot fitted with rubber bung bored to take carbon cathode and L-shaped outlet tube, as shown in Figure 7.5
Lead anode
Filtration apparatus
Crystallizing dish
Supply of sulphur dioxide gas
12 volt d.c. supply
Sodium metavanadate
1 M sulphuric acid
2 M sodium hydroxide

Procedure

Dissolve about 3 g of sodium metavanadate in 40 cm^3 of the sodium hydroxide solution in the beaker. Add 40 cm^3 of dilute sulphuric acid and then sufficient distilled water almost to fill the beaker. Transfer the solution to the Drechsel bottle, place in the fume cupboard, and pass sulphur dioxide through the solution until its colour has changed completely from yellow to blue.

Boil the solution until it no longer smells of sulphur dioxide. Cool the solution and assemble the apparatus shown in Figure 7.5. Electrolyse for thirty minutes when further reduction of the vanadium occurs.

Filter the contents of the cathode compartment into a crystallizing dish, cover with a piece of filter paper, and leave it for crystals to be formed.

Discussion

It is possible to achieve the second stage of this reduction by conventional chemical methods. From a table of $E^\ominus$ values (see pages 61–2) explain why sul-

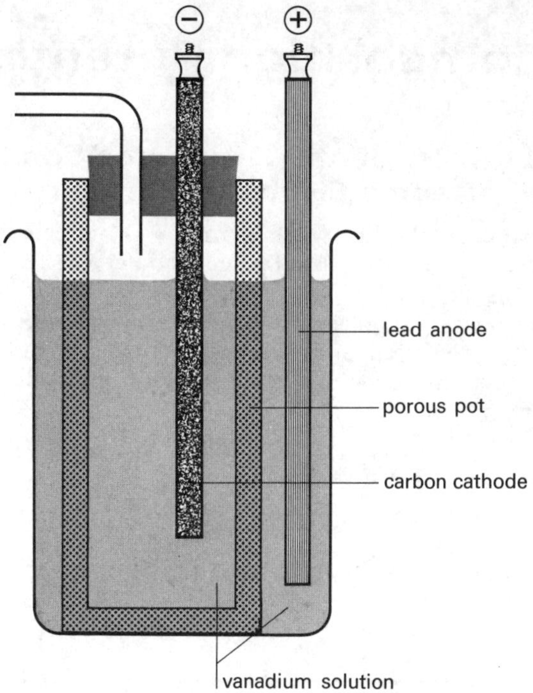

Figure 7.5 Electrolytic preparation of vanadium alum

phur dioxide reduces $VO_3^-(aq)$ only to the $VO^{2+}(aq)$ stage. What reducing agents could be used to bring about the further reduction? You have been asked to state the changes in oxidation number which occur by inspection of the equations. How would it be possible to find out the change in oxidation number which occurs in the last stage of the experiment by a modification of the electrolytic procedure?

Is there any reason why the whole of this reduction process should not be carried out electrolytically instead of the first part being done by using sulphur dioxide?

8: Decomposition Potentials

8.1 Investigation of the Decomposition Potentials of some Electrolytes

You might expect, if the theory of ionization is correct, that electrolysis would occur even if only a very small potential difference existed between the electrodes. Is this the case?

Devise a method of applying a continuously variable voltage up to 3 volts to an electrolytic cell, and for measuring the current flowing through the cell. After you have done this, compare your circuit diagram with that given in Figure 8.1.

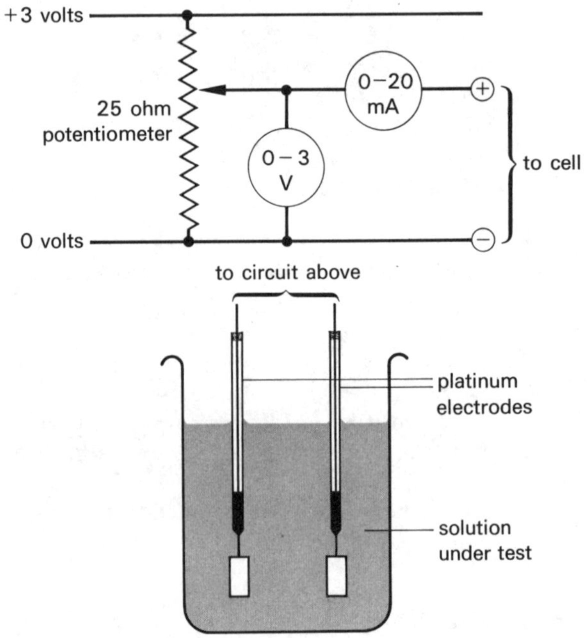

Figure 8.1 Determining some decomposition potentials

Apparatus and Materials

3 volt d.c. supply (this must be steady)
0–3 volt high resistance voltmeter
0–20 milliammeter
25 ohm potentiometer
2 platinum electrodes, with suitable holder
100 cm³ beaker
1 M solutions of sulphuric acid, hydrochloric acid, sodium hydroxide, and sodium chloride
0.05 M solution of hydrochloric acid

Procedure

Set up the apparatus shown in Figure 8.1. Using each electrolyte in turn obtain corresponding values for the current flowing (in mA) and the potential difference (in volts) for voltage increments of 0.25. Plot a graph of your results preferably using a different colour for the curves for each different electrolyte. The decomposition potential is obtained from the graph as shown in Figure 8.2.

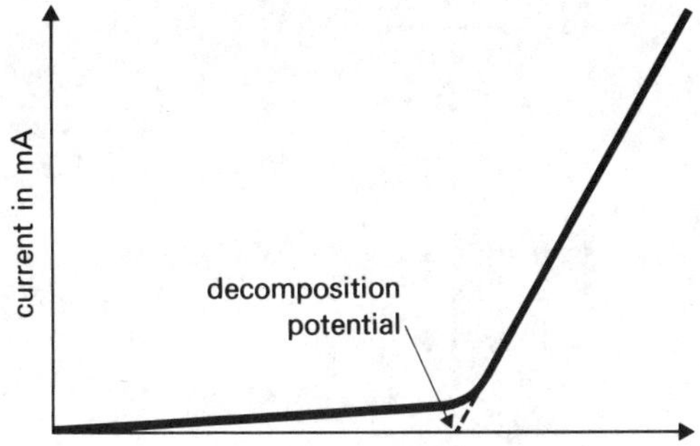

Figure 8.2

Discussion

Explain the values obtained for the decomposition potentials. Is the decomposition potential the same as electrode potential? If not, why?

Is the value of the decomposition potential for 1 M hydrochloric acid the same as that for 0.05 M acid? Attempt an explanation.

Why is it that in the electrolysis of copper(II) sulphate between copper electrodes a very small applied potential is sufficient to start the process, but after a time an appreciably higher potential has to be applied?

8.2 Does the Decomposition Potential of an Electrolyte depend on the Nature of the Electrodes?

The Castner–Kellner cell for the industrial production of sodium hydroxide from sodium chloride solution uses a mercury cathode. This experiment will help you to find out why.

Apparatus and Materials

As for Experiment 8.1, but with the electrolytic cell shown in Figure 8.3
2 volt d.c. supply
Boiling tube
Universal indicator paper

Procedure

Set up the apparatus as shown in Figure 8.1 but using the alternative cell shown in Figure 8.3. Find the decomposition potential for sodium chloride solution graphically as in the last experiment. Connect the cell directly to a 2 volt supply and electrolyse the solution for twenty minutes. Run the mercury off into a boiling tube containing a little distilled water. Try to identify any gas evolved.

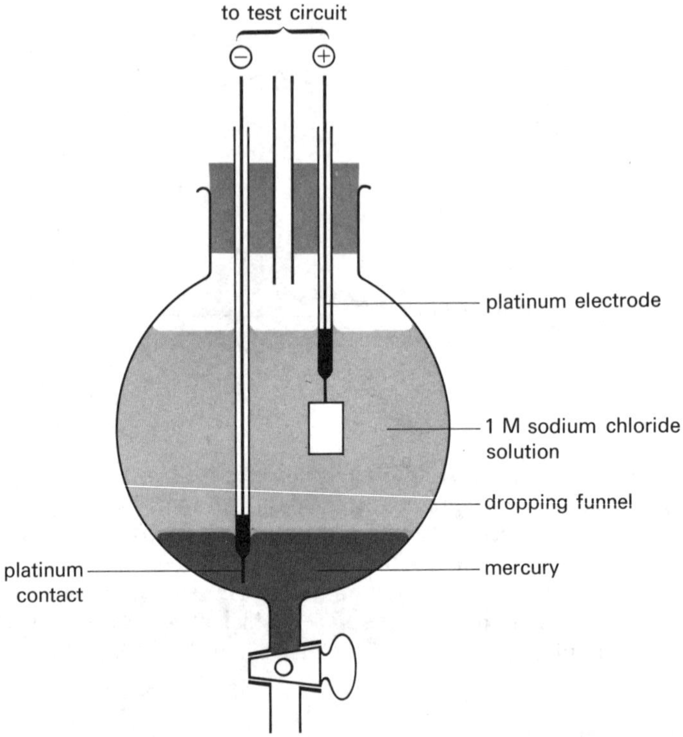

Figure 8.3 Effect of mercury cathode on the decomposition potential of sodium chloride solution

Test the liquid above the mercury with universal indicator paper. When the reaction between the mercury layer and the aqueous layer is complete separate the mercury, carefully dry it, and return it to the stock bottle.

8.3 Use of Decomposition Potentials in Analysis

In Experiment 8.1 you have established the fact that the decomposition potentials of solutions of different ions are different. Use is made of this in a method of electrochemical analysis known as *polarography*. If a solution containing a mixture of metallic ions has a gradually increasing potential applied to electrodes placed in it, a small current will flow at first, as was observed in Experiment 8.1. As the potential is increased the current suddenly rises steeply as the discharge potential of one of the ions is reached. The current then levels off again because of 'concentration polarization'. The explanation of this is that the ions in the immediate neighbourhood of the cathode are discharged first and they have to be replaced by others diffusing into the cathode region from the bulk of the solution. The diffusion is slow, so the increase in current is retarded (refer back here to your answer to the question about the electrolysis of copper(II) sulphate solution at the end of Experiment 8.1). If one metal ion species is present a curve of the type shown in Figure 8.4 is obtained.

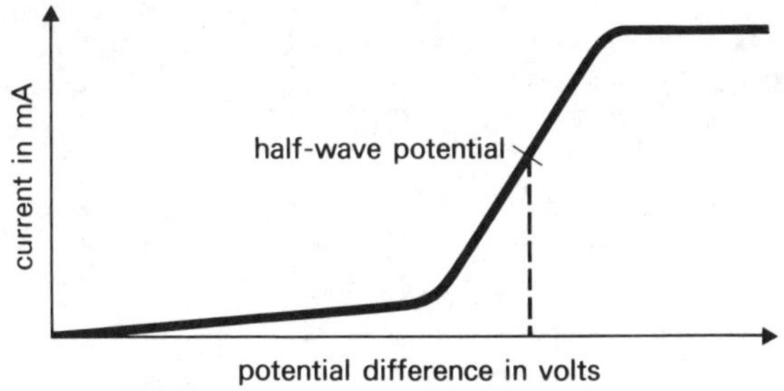

Figure 8.4

The speed at which this concentration polarization sets in, and hence the height of the 'plateau' in the curve, is dependent on the concentration of the ions. Further, the potential half way up the rise (or 'wave' as it is called, although not strictly a wave) is characteristic of a particular ion. It is called the half-wave potential'. If a solution contains more than one metal ion after one of them has been discharged, further ions will discharge as their decomposition potential is reached, and thus the current–voltage curve of a solution containing a mixture of ions will show a number of steps (or 'waves') each corresponding to the discharge of a particular ion species. Comparison of the half-wave potentials with standard tables enables us to identify the ions present. Further, by using standard solutions of ions and comparing the heights of the 'waves' obtained with those given by the mixture we can determine approximately the concentration of each ion present.

The instrument is called a 'polarograph', and uses a mercury anode and a dropping mercury cathode. Find out why.

In this experiment we shall find the half-wave potentials for cadmium and nickel. You might then try your hand at finding the constituents of a mixture.

Apparatus and Materials

25 ohm potentiometer
8.2 ohm fixed resistor
Sensitive galvanometer
Valve voltmeter or high resistance voltmeter (0–1.5 volts)
2 volt accumulator
Connecting wires
Polarography cell as illustrated in Figure 8.5
10^{-4} M solutions of cadmium chloride and nickel(II) chloride made up in 0.1 M potassium chloride
0.2 per cent aqueous solution of gelatin
Nitrogen cylinder and control gear

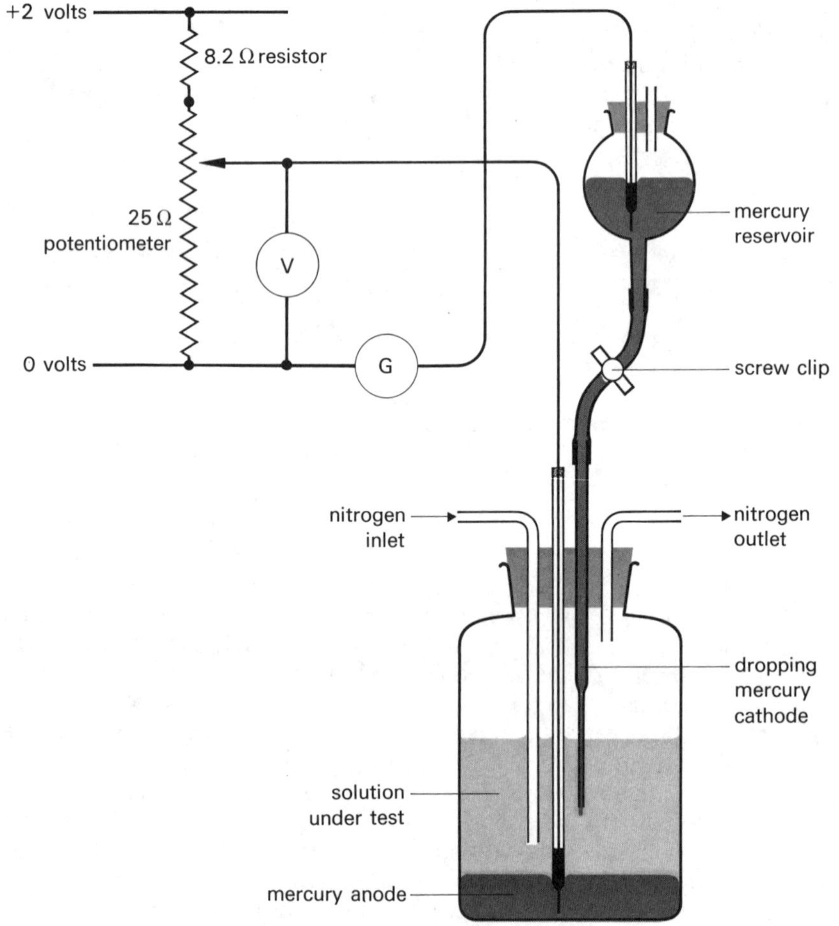

Figure 8.5 Apparatus for polarography

Procedure

Place 50–100 cm^3 cadmium chloride solution in the polarography cell and add five drops of the gelatin solution. (This solution is known as a maximum suppressor and gives a more manageable polarogram. How do you think it works?) Pass nitrogen through the cell for two minutes to displace dissolved oxygen and adjust the dropping rate for the mercury cathode to one drop every five seconds with the voltmeter reading 0.5 volt. Switch the galvanometer to minimum sensitivity, set the voltmeter to maximum reading, and adjust the galvanometer to full-scale deflection. This procedure is to ensure a suitable drop rate (which is voltage sensitive) over the voltage range employed.

Now set the voltmeter to zero and record the galvanometer reading. Increasing the voltmeter reading by 0.25 volt increments, obtain a series of corresponding readings for voltage and current. Plot your results on a curve of current against potential difference. (It is not necessary to assign particular units to the current readings.) Find the half-wave potential for cadmium. Repeat the experiment using the nickel(II) chloride solution.

Now try to find the constituents of the mixture of ions provided. If necessary, use the table of half-wave potentials given on page 62.

9: Conductivity Measurements

The Measurement of the Resistance and Conductivity of Electrolyte Solutions

The next part of our work will be concerned with the determination of the resistance and conductivity of electrolytes and the information that can be obtained from these measurements. Before describing the experiments themselves it is necessary to explain how the resistance of an electrolyte can be found and to discuss some definitions.

The Wheatstone Bridge and the Kohlrausch Bridge

The Wheatstone Bridge method of determining resistance depends on the following principle. In the circuit shown in Figure 9.1, with the hypothetical values for voltage and current indicated, if the galvanometer

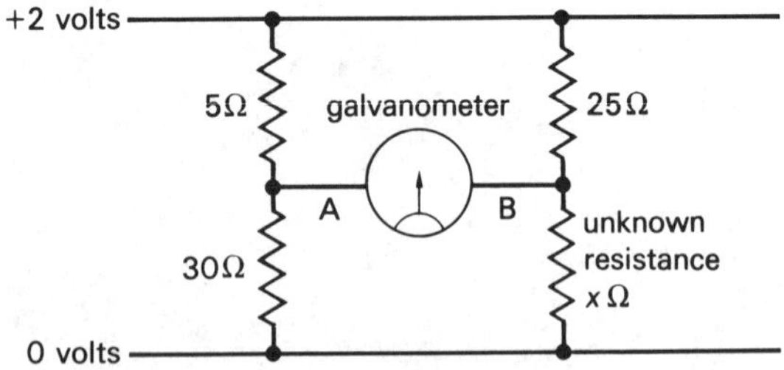

Figure 9.1

reading is zero there will be no current flowing through it and the potential difference between its terminals must be zero. Hence the potentials at A and B must be zero. For this to be the case the potential drop down the 5 ohm resistance must be the same as that down the 25 ohm resistance, and that down the 30 ohm resistance must be the same as that down the unknown resistance x. It follows that

$$\frac{5}{30} = \frac{25}{x}$$

or

$$x = 150 \text{ ohms}$$

This arrangement is readily used to find the resistance of solid conductors, and is very accurate indeed. For the determination of the resistance of electrolytes, however, some modification is necessary. If direct current is used, gaseous electrolysis products form on the electrodes, increasing the resistance, and invalidating the results. If an alternating supply is used with a frequency of about 1000 Hz, and with suitable electrodes, this problem can be overcome. An ordinary galvanometer will not respond to a.c. and some other detector is therefore necessary. The detector used originally was a pair of low-impedance earphones and the null-point was indicated by the least intensity of sound. It was never possible to make the sound disappear altogether and the determination of a null-point under these conditions is therefore not very accurate. With the invention of solid-state diodes, however, modern instruments containing a direct current meter connected in a rectification circuit are available. The original work on this branch of electrochemistry was done by Kohlrausch and this apparatus is known as a Kohlrausch Bridge.

The resistance R of a uniform resistor is proportional to its length l and inversely proportional to its area of cross-section A.

$$R = \frac{\rho l}{A}$$

The proportionality factor, ρ, is called the resistivity of the material, and $1/\rho$ is the conductivity. The usual symbol for conductivity is σ. In electrochemistry we are more often concerned with the conductivity than with the resistivity of an electrolyte.

Consider a uniform cube of material of 1 cm side. The length and area of cross-section of this cube are both unity. Hence the measured resistance of such a cube of material would be ρ, and this is how ρ can be defined. It is the resistance between opposite faces of a centimetre cube of the substance and its units are therefore Ω cm. The conductivity, being the reciprocal of the resistivity has units Ω^{-1} cm^{-1}.

It is less easy to define such a cube in the case of an electrolyte than it is for a solid conductor, and in practice no attempt is made to make the electrodes of a conductivity cell exactly 1 cm^2 in area and 1 cm apart. It is more convenient to determine for each particular cell a constant, known as the cell constant, which when multiplied by the measured conductance gives the conductivity:

$$\text{conductivity} = \text{measured conductance} \times \text{cell constant}$$

The cell constant is found experimentally by filling the cell with 0.02 M potassium chloride and finding its resistance. At 18 °C this solution is known to have a conductivity of 0.0024 Ω^{-1} cm^{-1}. Conductivity cells or pairs of electrodes are normally supplied with the cell constant marked. The platinum black on the surface of the platinum electrodes helps to prevent polarization by catalyzing the reverse of those reactions in which gases are formed.

9.1 Investigation of the Changes in Conductivity that take place during Acid–Base Titrations and Use of the Technique to find the Ethanoic (Acetic) Acid Content of Vinegar

You have already studied experimentally the changes of pH which occur in acid–base titrations. You would expect analogous changes to take place in conductivity. In this experiment you will find out if this is the case, and will use the technique of conductometric titration to find the concentration of a coloured acid solution, a determination not readily carried out by other methods.

Apparatus and Materials

Conductivity meter and dip electrodes
Magnetic stirrer and 'slave'
25 cm^3 pipette
250 cm^3 standard flask
10 cm^3 burette
50 cm^3 pipette (or measuring cylinder)
100 cm^3 beaker
0.1 M solution of hydrochloric acid
1.0 M solution of sodium hydroxide
Vinegar

(a) Conductivity changes when hydrochloric acid is titrated against sodium hydroxide

Procedure

Set up the apparatus shown in Figure 9.2. Observe the conductance of the 0.1 M hydrochloric acid. Run in 10 cm^3 of the sodium hydroxide in 1 cm^3 portions and record the conductance after each addition.

Results

Enter the results in a table as follows.

Volume of 1 M NaOH added in cm^3	*Observed conductance in* Ω *cm*	*Conductivity in* $\Omega^{-1}\ cm^{-1}$

Plot a graph between conductivity and volume of sodium hydroxide solution added.

Discussion

Explain the shape of the graph you have obtained.

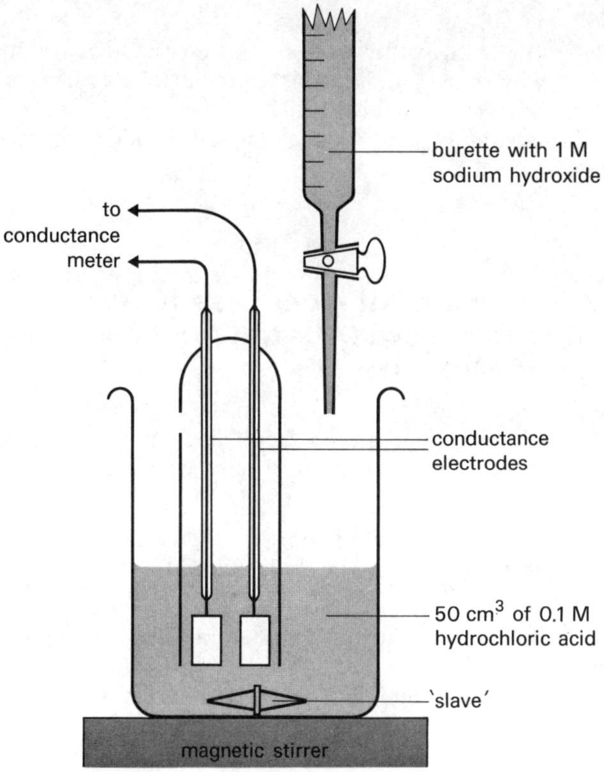

Figure 9.2 Conductance changes during titration

(b) Determination of ethanoic acid content of vinegar

Procedure

Pipette 25 cm^3 of vinegar into the 250 cm^3 measuring flask and make up to the mark with distilled water. Using 50 cm^3 of this solution instead of hydrochloric acid repeat the first part of the experiment.

Results

Draw up a table of results as in part (a) and plot a graph of conductivity against volume of sodium hydroxide solution added. From the graph calculate the volume of sodium hydroxide required to neutralize the acid in the vinegar, and calculate the concentration of the acid assuming it to be entirely ethanoic acid.

Discussion

1. Is the shape of the curve for vinegar the same as that for hydrochloric acid? If not, why? Can you link this up with the pH changes that occur in the neutralization of ethanoic acid?
2. What are the advantages and disadvantages of conductometric titrations?

9.2 Determination of the Dissociation Constant of Ethanoic (Acetic) Acid by Conductivity Measurements

We shall be finding the effect of dilution on the conductivity of 1.0 M solutions of ethanoic acid, hydrochloric acid, sodium chloride, and sodium ethanoate and for the calculation of our results we shall need to define three new quantities.

(1) *Dilution* V. This is the volume in dm^3 which contains one mole of solute. It is therefore the reciprocal of the molarity of the solution. Its units are $dm^3\ mol^{-1}$.

(2) *Degree of dissociation* α. This is the fraction of the total number of molecules of a weak electrolyte that are dissociated into ions. Thus, if we started with one mole of a weak acid HA, and when this was dissolved in water 0.25 mole of $H^+(aq)$ ions were formed, the degree of dissociation would be 0.25.

(3) *Molar conductivity* Λ. This is the product of the conductivity of the solution, σ, and the dilution.

$$\Lambda = \sigma V$$

As the conductivity of a solution is dependent on the concentration of ions in the solution it should be possible to find this concentration and hence the degree of dissociation by conductivity measurements.

Apparatus and Materials

Conductance meter and dip electrodes
Magnetic stirrer and 'slave'
500 cm^3 beaker
250 cm^3 measuring cylinder
50 cm^3 burette
Deionized water
1.0 M solutions of ethanoic acid, hydrochloric acid, sodium chloride, and sodium ethanoate

Procedure

Put 250 cm^3 of deionized water into the beaker, fill the burette with 1.0 M hydrochloric acid, and complete the assembly of the apparatus as shown in Figure 9.3. Find the conductance of the deionized water. This should be

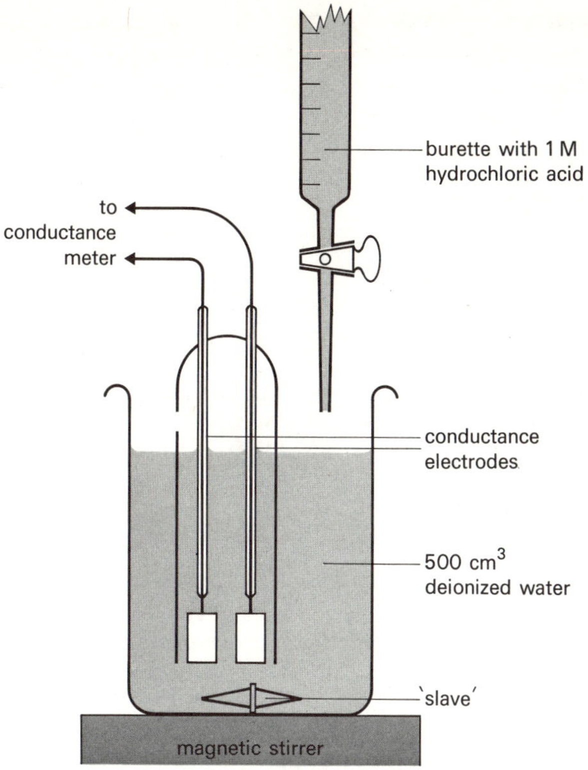

Figure 9.3 Dissociation constant for ethanoic acid by conductance

very low. Run in the hydrochloric acid in 0.5 cm^3 portions until 5 cm^3 in total have been added, and record the conductance after each addition. Then add the acid in 1 cm^3 portions until a total of 15 cm^3 has been added, and then in 5 cm^3 portions until the total volume added is 50 cm^3. In each case record the conductance after each addition.

Repeat the experiment for the other three electrolytes.

Results

Record your results in tables such as the following (a separate table for each electrolyte).

Volume of 1.0 M HCl added in cm^3	*Dilution (V), in dm^3 mol^{-1}*	*Conductance in Ω cm*	*Conductivity in Ω^{-1} cm^{-1}*	*Molar conductivity* (Λ)

Plot the results on a graph as shown below (it is not essential that the origin should appear on either axis).

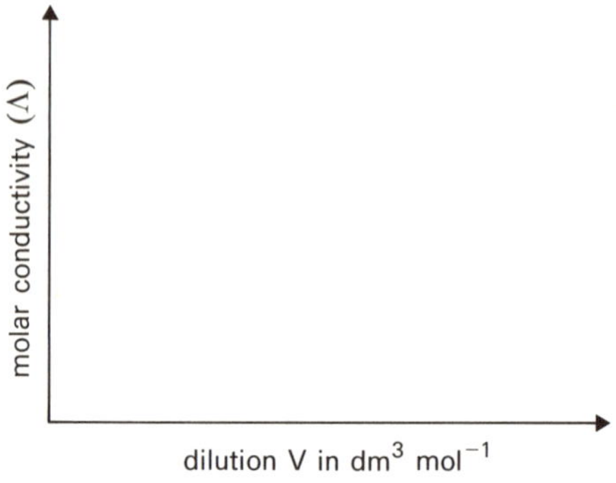

Figure 9.4

You will need separate graphs for each electrolyte the scale being chosen appropriately in each case.

Discussion

1. The conductivity (σ) is likely to be proportional to the number of ions in a given volume of the electrolyte. Thus, for *strong electrolytes* (i.e. those which are completely dissociated), because the number of ions in a given volume halves when the dilution is doubled, the molar conductance and hence the molar conductivity should be independent of dilution. Is this true for the strong electrolytes you have investigated? If there is any deviation suggest an explanation for it.

2. In the case of *weak electrolytes* the relationship between dilution and the number of ions present is not so simple, but it can be obtained by the use of the so-called Ostwald Dilution Law, which is simply the application of the equilibrium law to electrolytes. Thus, in the weak acid HA(aq) there is an equilibrium between undissociated HA, H^+, and A^-.

	$HA(aq)$ $\rightleftharpoons$	$H^+(aq)$ +	$A^-(aq)$
Concentration of each species if no dissociation took place (mol dm^{-3})	$\frac{1}{V}$	0	0
Concentration of each species when degree of dissociation is α (mol dm^{-3})	$\frac{1-\alpha}{V}$	$\frac{\alpha}{V}$	$\frac{\alpha}{V}$

$$K_c = \frac{[H^+(aq)]_{eqm}[A^-(aq)]_{eqm}}{[HA(aq)]_{eqm}} = \frac{(\alpha/V)^2}{(1-\alpha)/V} = \frac{\alpha^2}{V(1-\alpha)}$$

For electrolytes as weak as ethanoic acid, α is small compared with 1 and hence, to a good approximation,

$$K_c = \frac{\alpha^2}{V}$$

Hence, the fraction of the total number of molecules that is dissociated into ions in the case of ethanoic acid is proportional to $V^{1/2}$. Is your graph for the conductivity of ethanoic acid consistent with this conclusion?

Determination of the Degree of Dissociation for Ethanoic Acid

As you will observe, Ostwald's Dilution Law can be applied provided we have a value of α at a fixed dilution. Assuming that an electrolyte would be completely dissociated at infinite dilution, Arrhenius proposed the following relationship between α, the molar conductivity at a dilution *V*, and the molar conductivity at infinite dilution

$$\alpha = \frac{\Lambda_V}{\Lambda_\infty}$$

The molar conductivity at infinite dilution is the limiting value for molar conductivity as dilution increases. In the case of strong electrolytes it is readily possible to obtain this value from the molar conductivity/dilution curve. With weak electrolytes, however, such as ethanoic acid, no maximum is reached even when the solutions are so dilute that we cannot readily measure their conductivities. We can overcome this difficulty by applying Kohlrausch's Law of Independent Migration of Ions. This states that at infinite dilution the anions and cations in the electrolyte travel independently of each other, and so the molar conductivity at infinite dilution can be considered to be the sum of the molar conductivities due to the separate ions. If λ represents the molar conductivity of an individual ion

$$\Lambda_\infty^{HCl} = \lambda_\infty^{H^+} + \lambda_\infty^{Cl^-}$$

$$\Lambda_\infty^{CH_3CO_2Na} = \lambda_\infty^{Na^+} + \lambda_\infty^{CH_3CO_2^-}$$

$$\Lambda_\infty^{NaCl} = \lambda_\infty^{Na^+} + \lambda_\infty^{Cl^-}$$

$$\begin{aligned}\Lambda_\infty^{HCl} + \Lambda_\infty^{CH_3CO_2Na} - \Lambda_\infty^{NaCl} &= \lambda_\infty^{H^+} + \lambda_\infty^{Cl^-} + \lambda_\infty^{Na^+} + \lambda_\infty^{CH_3CO_2^-} - \lambda_\infty^{Na^+} - \lambda_\infty^{Cl^-} \\ &= \lambda_\infty^{H^+} + \lambda_\infty^{CH_3CO_2^-} \\ &= \Lambda_\infty^{CH_3CO_2H}\end{aligned}$$

Use your results for the molar conductivities at infinite dilution for hydrochloric acid, sodium ethanoate, and sodium chloride to calculate a value for the molar conductivity at infinite dilution of ethanoic acid. You can now use this to find α for ethanoic acid at various dilutions, and hence obtain values for K_c. Alternatively a computer can be programmed to calculate the average value of K_c from all the results.

How does your value for K_c agree with that obtained in Experiment 5.1 ?

Appendix

Some Standard Electrode Potentials

	$E^{\ominus}$ *in volts*
$Li^+ + e^- \longrightarrow Li$	−3.02
$K^+ + e^- \longrightarrow K$	−2.92
$Ca^2 + 2e^- \longrightarrow Ca$	−2.76
$Na^+ + e^- \longrightarrow Na$	−2.71
$Mg^2 + 2e^- \longrightarrow Mg$	−2.40
$Al(OH)_4^- + 3e^- \longrightarrow Al + 4OH^-$	−2.35
$Al^{3+} + 3e^- \longrightarrow Al$	−1.66
$Zn(OH)_4^{2-} + 2e^- \longrightarrow Zn + 4OH^-$	−1.22
$V^{2+} + 2e^- \longrightarrow V$	−1.20
$Mn^{2+} + 2e^- \longrightarrow Mn$	−1.10
$Te + 2e^- \longrightarrow Te^{2-}$	−0.91
$SO_4^{2-} + H_2O + 2e^- \longrightarrow SO_3^{2-} + 2OH^-$	−0.90
$Se + 2e^- \longrightarrow Se^{2-}$	−0.77
$Zn^{2+} + 2e^- \longrightarrow Zn$	−0.76
$Cr^{3+} + 3e^- \longrightarrow Cr$	−0.51
$S + 2e^- \longrightarrow S^{2-}$	−0.51
$Fe^{2+} + 2e^- \longrightarrow Fe$	−0.44
$Cr^{3+} + e^- \longrightarrow Cr^{2+}$	−0.41
$Cd^{2+} + 2e^- \longrightarrow Cd$	−0.40
$Ni^{2+} + 2e^- \longrightarrow Ni$	−0.25
$V^{3+} + e^- \longrightarrow V^{2+}$	−0.25
$Sn^{2+} + 2e^- \longrightarrow Sn$	−0.14
$Pb^{2+} + 2e^- \longrightarrow Pb$	−0.13
$2H_3O^+ + 2e^- \longrightarrow H_2 + 2H_2O$	0.00
$Cu^{2+} + e^- \longrightarrow Cu^+$	+0.17
$SO_4^{2-} + 2H^+ + 2e^- \longrightarrow SO_3^{2-} + H_2O$	+0.17
$Sn^{4+} + 2e^- \longrightarrow Sn^{2+}$	+0.20
$Cu^{2+} + 2e^- \longrightarrow Cu$	+0.35
$VO^{2+} + 2H^+ + e^- \longrightarrow V^{3+}$	+0.36
$O_2 + 2H_2O + 4e^- \longrightarrow 4OH^-$	+0.40
$I_2 + 2e^- \longrightarrow 2I^-$	+0.58
$Fe^{3+} + e^- \longrightarrow Fe^{2+}$	+0.75

$Ag^+ + e^- \longrightarrow Ag$	+0.81
$Hg^{2+} + 2e^- \longrightarrow Hg$	+0.86
$2Hg^{2+} + 2e^- \longrightarrow Hg_2^{2+}$	+0.79
$NO_3^- + 4H_3O + 3e^- \longrightarrow NO + 6H_2O$	+0.95
$Br_2 + 2e^- \longrightarrow 2Br^-$	+1.07
$4H_3O^+ + O_2 + 4e^- \longrightarrow 6H_2O$	+1.24
$CrO_4^{2-} + 8H_3O^+ + 3e^- \longrightarrow Cr^{3+} + 12H_2O$	+1.30
$MnO_2 + 4H_3O^+ + 2e^- \longrightarrow Mn^{2+} + 6H_2O$	+1.35
$Cl_2 + 2e^- \longrightarrow 2Cl^-$	+1.36
$Au^{3+} + 3e^- \longrightarrow Au$	+1.50
$PbO_2 + 4H_3O^+ + 2e^- \longrightarrow Pb^{2+} + 6H_2O$	+1.44
$ClO_3^- + 6H_3O^+ + 6e^- \longrightarrow Cl^- + 9H_2O$	+1.44
$MnO_4^- + 8H_3O^+ + 5e^- \longrightarrow Mn^{2+} + 12H_2O$	+1.50
$Pt^2 + 2e^- \longrightarrow Pt$	+1.60
$MnO_4^- + 4H_3O^+ + 3e^- \longrightarrow MnO_2 + 6H_2O$	+1.63
$Co^{3+} + e^- \longrightarrow Co^{2+}$	+1.80
$Pb^{4+} + 2e^- \longrightarrow Pb^{2+}$	+1.80
$O_3 + 2H_3O^+ + 2e^- \longrightarrow O_2 + 3H_2O$	+1.90
$F_2 + 2e^- \longrightarrow 2F^-$	+2.85

Some Polarograph Half-wave Potentials

All values for a background medium of 0.1 M potassium chloride at 298 K.

Ion	*Half-wave potential in volts*
Cd^{2+}	−0.60
Ce^{2+}	−1.40
Cr^{3+}	−0.90
Cu^{2+}	+0.04
In^{3+}	−0.61
Ni^{2+}	−1.10
Pb^{2+}	−0.40
Zn^{2+}	−1.00

Activities of Some Electrolyte Solutions

Concentration in mol dm^{-3}	$AgNO_3$	$CuSO_4$
0.001	1.0	0.74
0.005	0.92	0.53
0.01	0.90	0.41
0.05	0.79	0.21
0.1	0.72	0.16

Values for other concentrations can be estimated by graphical interpolation.

The table above gives the activity coefficients for silver nitrate and copper sulphate solutions at typical concentrations. For accurate work the activity of the ion is found by multiplying the molarity of the solution by the activity coefficient:

e.g. Activity of 0.1 M $Ag^+(aq)$ in silver nitrate $= 0.1 \times 0.72$

$$a_{Ag^+} = 0.072$$